Economics and Utopia
Geoff Hodgson

Critical Realism in Economics
Edited by Steve Fleetwood

The New Economic Criticism
Edited by Martha Woodmansee and Mark Osteeen

What do Economists Know?
Edited by Robert F. Garnett, Jr.

Postmodernism, Economics and Knowledge
Edited by Stephen Cullenberg, Jack Amariglio and David F. Ruccio

The Values of Economics
An Aristotelian perspective
Irene van Staveren

How Economics Forgot History
The problem of historical specificity in social science
Geoffrey M. Hodgson

Intersubjectivity in Economics
Agents and structures
Edward Fullbrook

The World of Consumption, 2nd Edition
The material and cultural revisited
Ben Fine

Reorienting Economics
Tony Lawson

Toward a Feminist Philosophy of Economics
Edited by Drucilla K. Barker and Edith Kuiper

The Crisis in Economics
Edited by Edward Fullbrook

The Philosophy of Keynes' Economics
Probability, uncertainty and convention
Edited by Jochen Runde and Sohei Mizuhara

Postcolonialism Meets Economics
Edited by Eiman O. Zein-Elabdin and S. Charusheela

The Evolution of Institutional Economics
Agency, structure and Darwinism in American institutionalism
Geoffrey M. Hodgson

Transforming Economics
Perspectives on the critical realist project
Edited by Paul Lewis

New Departures in Marxian Theory
Edited by Stephen A. Resnick and Richard D. Wolff

Markets, Deliberation and Environment
John O'Neill

Markets, Deliberation and Environment

Markets, Deliberation and Environment offers a systematic examination and critique of market-based approaches to environmental policy. The book includes discussion of the ethical boundaries of markets, the role of private property rights in environmental protection, the nature of sustainability and the valuation of goods over time.

John O'Neill considers the strengths and weaknesses of the main deliberative responses to the failure of market-based approaches to environmental policy. He defends the need for a more deliberative approach, which recognises the problems with recent formal experiments in deliberative institutions.

This multidisciplinary book spans economics, philosophy, political theory and environmental studies and should prove to be essential reading for undergraduate and postgraduate students studying courses in ecological and environmental economics.

John O'Neill is Professor of Political Economy at Manchester University. He has written widely on the philosophy of economics, political theory, environmental policy, ethics and the philosophy of science.

Economics as Social Theory

Series edited by Tony Lawson
University of Cambridge

Social Theory is experiencing something of a revival within economics. Critical analyses of the particular nature of the subject matter of social studies and of the types of method, categories and modes of explanation that can legitimately be endorsed for the scientific study of social objects, are re-emerging. Economists are again addressing such issues as the relationship between agency and structure, between economy and the rest of society, and between the enquirer and the object of enquiry. There is a renewed interest in elaborating basic categories such as causation, competition, culture, discrimination, evolution, money, need, order, organisation, power probability, process, rationality, technology, time, truth, uncertainty, value etc.

The objective for this series is to facilitate this revival further. In contemporary economics the label 'theory' has been appropriated by a group that confines itself to largely asocial, ahistorical, mathematical 'modelling'. Economics as Social Theory thus reclaims the 'theory' label, offering a platform for alternative rigorous, but broader and more critical conceptions of theorising.

Other titles in this series include:

Economics and Language
Edited by Willie Henderson

Rationality, Institutions and Economic Methodology
Edited by Uskali Mäki, Bo Gustafsson and Christian Knudsen

New Directions in Economic Methodology
Edited by Roger Backhouse

Who Pays for the Kids?
Nancy Folbre

Rules and Choice in Economics
Viktor Vanberg

Beyond Rhetoric and Realism in Economics
Thomas A. Boylan and Paschal F. O'Gorman

Feminism, Objectivity and Economics
Julie A. Nelson

Economic Evolution
Jack J. Vromen

Economics and Reality
Tony Lawson

The Market
John O'Neill

Markets, Deliberation and Environment

John O'Neill

LONDON AND NEW YORK

First published 2007
by Routledge
2 Park Square, Milton Park, Abingdon, Oxon OX14 4RN

Simultaneously published in the USA and Canada
by Routledge
270 Madison Avenue, New York, NY 10016

Routledge is an imprint of the Taylor & Francis Group, an informa business

Typeset in Times
by Keystroke, 28 High Street, Tettenhall, Wolverhampton
Printed and bound in Great Britain
by TJ International Ltd, Padstow, Cornwall

British Library Cataloguing in Publication Data
A catalogue record for this book is available from the British Library

Library of Congress Cataloging in Publication Data
O'Neill, John, 1956–
Markets, deliberation, and environmental value / John O'Neill.
p. cm.
Includes bibliographical references and index.
1. Environmental policy–Economic aspects. 2. Environmental policy–Social aspects. 3. Public goods–Valuation. 4. Contingent valuation. 5. Environmental ethics. I. Title.
HC79.E5O514 2006
333.7—dc22
2006014465

ISBN10: 0–415–39711–1 (hbk)
ISBN10: 0–415–39712–X (pbk)

ISBN13: 978–0–415–39711–7 (hbk)
ISBN13: 978–0–415–39712–4 (pbk)

In memory of
Bill O'Neill
1917–2005

Contents

Acknowledgements

This book is the result of a number of research projects and collaborations in the United Kingdom and Europe since the early 1990s. There are a large number of people whose conversations and comments have informed my thinking on the issues I discuss here and I regret that I cannot name them all here. My greatest debt has been to Alan Holland. My conversations with him have been the starting point for a large number of the arguments in this book. I would also like to thank my other colleagues during my time at Lancaster University for their support in writing this book and their many comments and discussions on earlier drafts. My undergraduate and postgraduate students over the years have listened to and commented on previous versions of the chapters in this book: their fine critical comments were a reminder that teaching and research are not separate activities. Parts of the book were completed during a fruitful period as a Hallsworth Senior Research Fellow at Manchester University: I owe a particular debt of gratitude to Thomas Uebel, Hillel Steiner, Noel Castree and Pat Devine. I would like to thank Jacqui Burgess and her colleagues at University College London for their openness in giving me access to transcripts of the in-depth discussion groups on the Pevensey Levels contingent valuation and for their many valuable discussions. Jacqui Burgess, Tim O'Riordan and Andy Stirling were partners in a project on deliberative institutions and I gained much from my own deliberations with them. There are many other colleagues in the United Kingdom to whom I owe thanks for their conversations and comments on earlier versions of the chapters in this book. I would particularly like to thank Jonathan Aldred, Michael Banner, John Barry, Wilfred Beckerman, Ric Best, Bob Brecher, Claudia Carter, Graeme Chesters, Andrew Dobson, Tim Hayward, Michael Jacobs, Russell Keat, Sue Owens, Andrew Sayer, Sigrid Stagl, Kerry Turner and Jo Wolff. In the UK, earlier versions of chapters in this book were given to seminars and conferences at Aberdeen, Bristol, Cambridge, Durham, Edinburgh, Glasgow, Keele, Kings College London, Lancaster, Leeds, Norwich, Oxford and University College London. I would like to thank participants for the many helpful comments made on those occasions.

Many of the ideas of the book have been developed through collaborations in a number of European projects. I benefited a great deal from conversations with Clive Spash, who has been a central partner in many projects and a co-author with me of a report for the European Commission on conceptions of value in environmental decision making. Joan Martinez-Alier and Giuseppe Munda at the University of Barcelona co-authored a series of papers with me on value incommensurability and multi-criteria analysis in ecological economics. I learned much from the work we did together and from my many wide-ranging conversations with Joan Martinez-Alier, especially around issues on environmental inequalities. Martin O'Connor co-authored a report with Alan Holland and myself on the economic valuation of environmental damage – a terrific period of intense research which led to a number of further collaborations over the years. Arild Vatn was a partner in a number of projects and has been a constant source of encouragement and ideas that have been central to the themes in this book. I presented an earlier version of the book as an intensive course for research students in Oslo and I would like to thank both students and colleagues for their comments. Juan Sánchez García and Federico Aguilera Klink from La Laguna in Tenerife were also co-researchers in a number of projects. I learned much from them, not only about economics but also about how to properly appreciate life. I would also like to express particular thanks to Sybille van de Hove in Barcelona, Anton Leist and Peter Schaber in Zurich, Elisabeth Nemeth and Friedrich Stadler in Vienna, Silvio Funtowicz and Bruna de Marchi in Italy, Markku Oksanen, Yrjö Haila and Juha Heidenpa in Finland and Finn Arler in Denmark.

I have also benefited from a number of fruitful collaborations with colleagues outside of Europe. I owe a great deal to my many conversations with Avner de Shalit in Jerusalem and Andrew Light in the United States. Particular thanks are also owed to John Dryzek, Robyn Eckersley, Dale Jamieson, Bryan Norton and Val Plumwood. Earlier versions of chapters were given in seminars and conferences in Athens, Austin, Barcelona, Geneva, Helsinki, Jerusalem, Maine, Mainz, Mannheim, Melbourne, Odense, Oslo, Paris, Pori, Roskilde, Tampere, Tenerife, Tunisia, Turku, Vienna, Wageningen, Yale and Zurich. I would like to thank the participants for their critical comments on those occasions.

I would like to end with some more personal debts of gratitude. I would like to thank Yvette Solomon and our two children Bridie and Rosie for their constant support. I wouldn't have been able to complete this book without them. In my acknowledgements to my last book I began with thanks to two building workers who had a large influence on my thought and life, my father Bill O'Neill and my friend Joe Megechie. Both have died since I wrote that book. The influence of each is still to be found in this book. My father died just as I was finishing the final draft. I would like to dedicate this book to his memory.

Introduction

Globalisation and environment

Who is against globalisation?

An apparent paradox in debates around globalisation is that, certain nationalist and fundamentalist groups aside, it is hard to find many voices against globalisation as such. Most of those in what is called the 'anti-globalisation' movement are happy to defend particular forms of globalisation and engage in a variety of global actions. Hence, for example, the development of global activist networks, of regional and world social fora, of alliances such as 'People's Global Action', and of slogans such as 'globalisation from below'. At the same time, the apparent paradox also runs in the opposite direction: many who are taken to be the central proponents and architects of globalisation are often against the development of certain forms of globalisation, for example of global legal structures. However, the paradoxes are not deep ones. That there are few unequivocal voices for or against globalisation as such should not be surprising. The term 'globalisation' unqualified by an adjective or noun is of itself empty. The question needs to be answered, 'the globalisation of what?' Debates around globalisation include at least the following: the globalisation of particular economic structures and institutions; cultural globalisation; the globalisation of structures of governance; the globalisation of law; the globalisation of civil society; the globalisation of technology; the globalisation of communication systems; the globalisation of knowledge; the globalisation of ethical and political discourse. These different dimensions of globalisation can themselves take different forms – in particular, there are a variety of forms that economic globalisation could take other than the particular neo-liberal form that now predominates. These different dimensions and forms of globalisation can stand in variety of different relations with each other.

The central target of most anti-globalisation activism is a particular form of economic globalisation, neo-liberalism, and the global institutions that are taken to be the principal agents for the realisation of that form of globalisation such as the World Trade Organization (WTO), International Monetary Fund (IMF) and the World Bank. A defining feature of neo-liberalism is the project for the globalisation of markets. The project for such globalisation is justified by a variety of arguments that draw on competing neoclassical and Austrian traditions in economics and particular forms of liberal political theory. While those traditions differ both in their normative and empirical assumptions and in their conclusions about the proper scope of market mechanisms, they have been the intellectual sources of defence of market globalisation.

What is meant by market globalisation? There are two distinct senses in which one can talk of globalisation of markets in this context. First, the term can refer to the expansion in the geographical and spatial scope and penetration of markets. The globalisation of markets in this sense is manifested in a variety of ways: in the quantitative growth in the extent and intensity of global trade relations; in the deregulation of market relationships; in the weakening of the capacity of national and regional bodies to govern market transactions, including for example the capacity to implement social and environmental standards that are taken to represent barriers to trade; in the expansion of market relations into marginal communities that previously relied upon non-market economic relations, such as common property regimes, mutual aid and gift. Second, the term can refer to expansion of the domain of goods and spheres of activity governed by market norms, that is with the spread of market norms and relations into spheres in which they were previously absent. A number of goods and services have remained outside the sphere of the market or are such that their status as commodities which could be bought and sold on markets is contested. There are a variety of different such goods, but they have included at least the following: certain publicly provided goods which are often taken to be conditions of basic human flourishing, such as water, sanitation, health services and essential environmental services; public goods such as education and knowledge; items that are constitutive of personal integrity such as bodily parts and reproductive services; goods that are constitutive of political community, such as votes, citizenship rights, political office and access to political power; goods that are constitutive of the social relations of particular communities, such as shared public spaces, shared cultural goods, gifts that foster social solidarity. Many such non-market domains have become increasingly colonised by markets.

There are two distinct ways in which market norms and relations might expand into these new spheres. First, they can expand directly. Items that were previously outside the market are transformed into commodities subject to market exchange. Second, they can also expand indirectly where the relations, attitudes, forms of evaluation and norms characteristic of the market are transferred to non-market spheres. Both forms of colonisation can be illustrated in recent developments in higher education. At present higher education remains predominantly a non-market good that is publicly provided. Its status as such is increasingly threatened by the attempt to include educational services in the discussions of the General Agreement on Trade in Services (GATS) which would transform educational services directly into tradable commodities. However, the indirect expansion of market norms and forms of valuation has taken place already prior to the projected direct expansions of market norms and relations into higher education. For example, while in the United Kingdom, higher education 'services' are predominantly not yet items that are directly traded on a free market, there has been for some time an increasing replication of the language and relations of the market, for example as students are redefined as 'consumers' of a service, and research and teaching become the objects of managerial audits to be appraised by market-based modes of evaluation.

Much of the resistance to globalisation that forms the anti-globalisation movement has been a resistance to the specific neo-liberal project of globalising markets. The movement has involved resistance to both the spatial expansion of markets and their expansion in scope. One has seen the defence of surviving common property regimes, gift economies and the public provision of goods against the incursion of markets. Opposition to the neo-liberalism is often expressed through opposition to direct the expansion of the sphere of the market, for example into water services, health provision, genetic goods, knowledge and educational spheres. It is also expressed through the development of new non-proprietary regimes in previously predominately commercial spheres, for example in the development of copyleft regimes, first in free software and subsequently in other cultural spheres. Indirect forms of market colonisation also form an object of criticism of those opposed to market globalisation. For example the corporate branding of cultural and educational goods and services may not turn either into tradable commodities – indeed they may allow them to be free at the point of use – but can be seen to corrupt the internal goods and standards of evaluation proper to cultural and educational practices.

Global civil society is often taken to be the central sphere of resistance to market globalisation. However, approaches to the processes of market globalisation projects within global civil society are clearly not uniform. There are tensions between different responses to the expansion of markets. One major tension is between those who take the existence of globalising markets as a social given and aim to shift the terms on which it takes place to defend particular goods and communities and those who are concerned principally to resist the very process of expanding markets. Consider for example the practice of what is often called 'bioprospecting' or 'biopiracy', the attempt to establish commercial intellectual property rights over particular genetic resources and chemical compounds found in plants and other biological entities. One response is to take the new intellectual property rights regime to be a social reality that actors have to live with and respond to. Thus for example one response to biopiracy that involves the appropriation of the resources and knowledge of marginalised 'indigenous' communities has been to attempt to define local property rights on those goods, thus bringing communities into the realm of international commerce but on more favourable terms. The private appropriation of such resources has to take place in ways that those communities are able to benefit from the process of commercialisation. Again a standard response to bioprospecting by public research bodies and universities previously committed to the development of knowledge as a publicly accessible good has been to establish their own intellectual property rights over genetic goods or to enter into partnerships with private corporations. However, another response to bioprospecting is to resist the very process of the extending commercial intellectual property rights over the goods in question and to defend genetic resources and knowledge about them as a global common good or public good. The global genetics commons the argument runs should not be subject to 'enclosure'. Genetic resources should not be treated as goods to be bought and sold in markets. The interests of the local communities whose knowledge and resources are threatened with appropriation by private corporations are not best met by merely shifting the terms of commercialisation. Public research needs itself to be kept free of commercialisation both for the good of the development of the knowledge as such, and in the interests of democracy that requires trustworthy sources of knowledge that are free of particular interests. Thus roughly runs some of the main lines of argument.

The conflicts here are typical of more general tensions within global civil society as to the appropriate response to the globalisation of market relations. Similar tensions are apparent in other spheres. On the one hand, certain parts of civil society rather than being oppositional to economic globalisation have become a part of that process. For example, with the retreat of the state as the primary agency for the delivery of welfare services and development policies, professional international non-government organisations (INGOs) are increasingly taking on such roles and become themselves primary vehicles for the delivery of policies of the agencies identified with neo-liberalism such as the World Bank. Again, INGOs in an attempt to defend particular goods in a commercial world have increasingly developed partnerships with large corporations. The development and sale of the WWF brand to the corporate world is perhaps the most prominent example. On the other hand global civil society contains networks and movements that stand in a more antagonistic relationship with market globalisation. Their activity is manifest not just in the protests surrounding the meetings of the WTO and the G8, but also in the development of different social fora and modes of public deliberation outside of state and market. Different sectors of global civil society have had then a variety of different relations to the processes of economic globalisation, some involved within those processes, some antagonistic to them.

There is a more general point to be made here about the conflicts and tensions between different processes of globalisation. These different relations between global civil society and economic globalisation are typical of the variety of different and conflicting relations that different forms and dimensions of globalisation can take to each other. Consider for example the globalisation of knowledge. The relation of neo-liberal economic globalisation and the globalisation of knowledge is a similarly complex one. The project for the globalisation of knowledge is one that goes back to the start of modern science. Science was to become, as Bishop Sprat put it in his *History of the Royal Society*, the 'Philosophy of Mankind':

> It is to be noted that they have freely admitted Men of different Religions, Counties, and Professions of Life. This they were oblig'd to do, or else they would come far short of the Largeness of their own Declarations. For they openly profess, not to lay the Foundation of an *English*, *Scotch*, *Irish*, *Popish*, or *Protestant* Philosophy; but a Philosophy of *Mankind*.
>
> (Bishop Sprat, *History of the Royal Society of London*, 1667, pp. 62–3, emphasis in the original)

The global aspiration for science was central to the cosmopolitan ambitions of the enlightenment. The scientific community was to be a community of humankind, a global community governed by norms of reason. For a scientific claim to be objective it must make a claim on any rational agent, independent of local cultural and social particularities. The language of science was to be a global lingua franca, scientific knowledge a global publicly accessible good. There are critics of this globalising enlightenment project. Those globalising ambitions are themselves understood as merely ways in which one local form of knowledge is exercised and imposed on others without power whose own knowledge is marginalised. Hence, for example, the defence of 'local' knowledge against the globalising ambitions of the enlightenment. Aimed against scientism, that is against the claim that the domain of science is coextensive with the domain of knowledge, the criticism does have power. There does exist knowledge of particulars, local to a particular time and place, that cannot be formulated in the general claims of science. There are forms of 'knowledge how', embodied in practice, that cannot be expressed in propositional form.

However, to criticise scientism is not to criticise science nor the globalising aspirations of the enlightenment. To place limits on the scope of abstract scientific knowledge is not to deny its own epistemic virtues as constitutive of a potential global community or that it makes truth claims that are objective in the sense of making claims on any rational agent. Moreover, the traditional enlightenment project for a global scientific community that defends a global publicly accessible good is in increasing conflict itself with the globalisation of the scope of markets. As we have noted above, the new regimes in intellectual property rights that extend commercial property rights to the domain of knowledge are inimical to the communal norms of publicity that have been constitutive of the scientific community and undermine the status of knowledge as a global public good. The traditional enlightenment model of globalised knowledge stands in conflict with the neo-liberal project for globalised markets.

The globalisation of moral and political discourse also stands in a complex set of relations to other forms and dimensions of globalisation. In the context of globalised political and economic relations it is often argued that there is a need for a global ethic. The existence of global economic and political interdependencies, and of globally shared problems, for example surrounding environmental problems such as global warming or biodiversity loss requires, the argument runs, a shared set of ethical concepts in which to frame international policy. The project of a cosmopolitan ethic that makes universal claims on all moral agents also reflects another of the globalising aspirations of the enlightenment. The emancipatory values that are part of an enlightenment inheritance are of particular appeal in this context. Emancipation requires a critical standpoint that transcends what is local. Local cultural practices are sometimes oppressive to particular individuals and groups, for example to women and subordinate castes and classes. To be able to develop criticism of those practices one needs a standpoint and set of normative concepts that transcends the local, that allows the possibility of standing outside particular social practices to formulate sceptical questions about them. The project requires an ethical discourse that is independent of particular cultures but rather specifies what is owed to any human being. Such an ethical discourse is often taken to require a minimalist vocabulary of 'thin' moral concepts, that employ general abstract terms of rights and goods that are taken to be universal, transcending the specific ethical understandings of local culture, in contrast with thick concepts whose understanding requires immersion in particular local practices. The language of human rights, for better or for worse, has come to embody that thin cosmopolitan ethic. However, it is not hard to find arguments that run in the opposite direction. The argument runs that in the context of the new globalisation, global normative discourse far from offering a standpoint to criticise illegitimate power and injustice embodied in local social practices, is becoming rather a way in which global economic and political power is itself expressed. The global ethical discourse that finds its way into international policy documents and treaties is employed by global elites to enforce a particular set of interests. Moreover, particular resistance to globalisation of markets is often articulated in terms of local moral meanings. They require geographically local terms – after all what is marketable in one place is not in others. They require terms that are specific to particular social spheres and practices – one cannot understand resistance to the extension of markets without reference to the social meaning of goods in particular cultural spheres.

This debate is one to which I will return later in Chapter 8. However, two comments are in order here. First, global conversation about values is inescapable. Radical relativism comes too early, before such conversations have begun, or too late, once they are under way. Second, arguments in defence of local cultural understandings implicitly call themselves upon the same emancipatory values of the enlightenment as its opponents – and they are none the worse for that. To draw attention to the ways in which particular voices and perceptions are silenced in global conversations draws its critical power on shared assumptions about the value of equality in standing, voice and power. What is I think true is that there are limits to any morally minimal language in articulating the limits of markets. It is certainly true that there are arguments within the tradition of enlightenment moral minimalism that are appealed to in limiting the scope of market relations. Consider for example Kant's distinction between a price and dignity:

> In the kingdom of ends everything has either a *price* or a *dignity*. If it has a price, something else can be put in its place as an equivalent; if it is exalted above all price and so admits of no equivalent, and then it has a dignity. What is relative to universal human inclinations and needs has a *market price* . . .; but that which constitutes the sole condition under which anything can be an end in itself has not merely a relative value – that is, a price – but has an intrinsic value – that is, *dignity*.
>
> (Kant, 1785, *Groundwork of the Metaphysics of Morals*, p. 77, emphasis in the original)

The position will take one some of the way in articulating the limits of markets, but it will not take us very far. There is very little that the Kantian account rules out. What comes under 'dignity' rules out the buying and selling of rational human agents – and that is clearly important. However, resistance to the globalisation of markets is not just about opposition to slavery. One might force more into the Kantian account by specifying the further conditions of human dignity. However, specifying what counts as an indignity looks as if it will require a much thicker set of ethical concepts. There is no escaping I think the need to appeal to a vocabulary that goes beyond a set of abstract minimal moral concepts. However, it is not clear to me for reasons I discuss later that an account of human flourishing we require to make universal moral claims requires the elimination of a thick moral language. A global ethic need not be a morally minimalist one.

Globalising markets and environmental goods

The arguments of this book can be understood against the background of these general debates about the globalisation of markets and its relation to other dimensions of globalisation. However, this is not a book that is about globalisation. It has a more modest focus. It is concerned with a particular class of goods, environmental goods. Environmental goods have been a key example of conflicts between proponents of the expansion of market norms and relations and those who are opposed to such expansion. Many environmental goods are currently non-market goods over which no property rights have been defined. As we shall see in Chapter 1, in standard economic theory this is seen as the source of our environmental problems. According to this view, the reason for the growing damage and destruction of environmental goods is that preferences for such goods are not registered in the market. The solution to the problems is to ensure that they are, either by directly extending property rights over environmental goods so that they can be traded as commodities on markets, or by indirectly extending prices over such goods by ascertaining what people would pay for them were there a market and employing those prices in standard cost-benefit analysis. The sphere of markets needs to be expanded to include environmental goods.

This view is one defended not just by liberal proponents of market solutions, but also by many everyday environmental practitioners faced with the loss of environmental goods. It is a view that runs into resistance. At a theoretical level it runs against a view that moves in exactly the opposite direction, that the source of our environmental problems lies not in the failure to expand market norms to all spheres, but in that very process of expansion. At a non-theoretical level it runs up against resistance to the process of putting market prices on goods for example by individuals who protest when asked their willingness to pay for some environmental good and communities who protest when asked their willingness to accept some sum of money for the loss of an environmental good. In Chapter 1 of this book, I defend the rationality of those responses against the view that our environmental problems require market solutions. On the one hand, refusal to put a price on certain environmental goods or to treat them as alienable property is constitutive of certain social relations and ethical commitments. To express those attitudes is not to express some irrational or strategically rational response but is to express proper concern for those goods. On the other hand, willingness to pay itself is reason-blind. It fails to register the reasons for a preference for some good, but only their intensity. Environmental decisions require deliberative fora rather than market based mechanisms.

The remaining chapters of Part I of the book develop different dimensions of this case against the view that the solution to our environmental problems lies in the expansion of market-based decision-making methods to include environmental goods. Chapter 2 considers some broadly utilitarian defences of cost-benefit analysis in the context of a particular policy-making context, the protection of biodiversity. Most sane economists do not think that economic decision-making tools such as cost-benefit analysis can replace the process of political deliberation. The contribution that the economist makes is taken to be more modest. Through the measurement and aggregation of preferences for different goods the economist is able to enter into the policy-making process with a recommendation of which option best improves total human welfare. It is for the political process to consider the place of that input about welfare with other political and ethical constraints on policy making for example around justice. I argue that this more modest view still fails in virtue of the account of well-being as preference satisfaction that it assumes. The process of deliberation is prior to an account of what makes for human well-being not simply subsequent upon it. Chapter 2 ends with consideration of still more modest pragmatic defences of monetary valuation as the best means to defend environmental ends in a policy-making context dominated by market norms. Against that position I argue that the development of more deliberative mechanisms for policy making matters not only as end in itself but also for the long-term realisation of environmental values.

Chapter 3 returns to a central claim in mainstream environmental economics, that to solve our environmental problems requires a set of tradable property rights to be defined over environmental goods. The monetary valuation of environmental goods, either hypothetical or actual, presupposes that those goods can be understood as commodities which could in principle be exchanged in markets. Chapter 3 gives some reasons for scepticism about this property rights solution by placing the arguments for the position in the context of an older influential tradition of arguments about property that goes back to Aristotle and which focus on the need for a distribution of care for particular goods. The modern defences of the extension of property rights to environmental goods often implicitly appeal to this tradition. Sustaining particular environmental goods requires the assignment of special relations of care towards them. However, while recent accounts often implicitly appeal to this tradition I show that they are in conflict with it. Care for particular goods that are constitutive of the life of a community within and between generations is often incompatible with the extension of tradable property rights over those goods. We express mutual ties of community through a resistance to the extension of exclusive property rights assigned to particular individuals or corporations. Moreover, conflicts between different accounts of what counts as care for a particular place are themselves expressed through conflicts around claims to property rights in particular goods. From Locke through to recent environmentalism, the appeal to particular accounts of care to justify property rights in particular goods has been used in very different ways to justify the appropriation by powerful social interests of the goods of marginalised social groups. Particular property rights regimes needs to be understood as ways in which particular distributions of power are developed and sustained.

Chapter 4 considers the view that environmental problems are best solved by bringing environmental goods directly into markets. There are arguments from within mainstream economics for scepticism about cost-benefit analysis as a mode of responding to environmental problems. In particular public choice traditions offer some reasons for scepticism. The argument runs that there is no reason to suppose that the assumptions concerning the self-interested motivations of actors employed in the explanation of market behaviour should cease to apply when we move to non-market spheres. In particular bureaucratic and political actors involved in the assessment of different projects should not be assumed to be benign actors concerned only with the public interest. Given self-interested bureaucrats and politicians pursuing their own interests in advancement and power, the 'market failures' responsible for environmental problems will be merely replaced with 'government failures'. The long and surprisingly uniform record of cost-benefit analyses of large public projects to underestimate costs and overestimate benefits might be seen as just such an example of what might happen. For public choice traditions in economics there is little reason for surprise in this tendency. The argument would run that the solution of environmental problems lies not in the surrogate prices and markets but rather in environmental goods and bads being themselves directly traded in actual markets. In Chapter 4 I suggest that, while some scepticism about benign motivations among state actors is appropriate, the public choice theory's assumptions about the universality of self-interested motivations in all institutional contexts should be rejected. The claim that the assumptions about the motivations of actors in the market sphere can be generalised to the explanation of behaviour in different spheres itself fosters a form of market globalisation. The differences between spheres and the norms and self-understandings that are constitutive of them are rendered invisible. In so far as there is a substantive claim being made by public choice theorists it is false. Different conceptions of an agent's interests are both defined by and fostered within different institutional settings. Individuals are motivated by a variety of different ends in different contexts, where some of these conflict with conceptions of self-interest fostered within the market. Correspondingly, the standard public choice approaches to social explanation and design on the assumption of universal avarice, should be replaced by those found in the older institutionalist tradition of economics which starts from the ways in which institutions define and foster different conceptions of interests. Chapter 4 outlines the implications of this institutionalist starting point for our understanding of environmental goods, arguing in particular for the significance of non-market and non-state associations in fostering environmental goods.

Part II considers the temporal dimension of the valuation of environmental goods. How do you value goods over time? In Chapters 5 and 6 I suggest that a particular market-based model of public policy making has crowded out the proper ethical and political considerations that should be at the centre of answers to that question. Chapter 5 considers the assumptions that underpin the cost-benefit approach to environmental decision making. Most of the discussion has focused on the assumption that the future should be discounted, that is that future benefits and costs count less than those in the present, the further into the future the less they count.

There are good reasons why this assumption has been at the centre of debate, since it appears to sanction intergenerational injustice. However, other assumptions in consequence often go unnoticed. Two in particular form the focus of Chapter 5. The first is that values of events at different moments of time are additively separable: the value of what happens at some point in time is independent of the value at what happens at any other point and the total value of a project over a period of time is the sum of these independent values. The second is that the past is irrelevant in considering the value of the project. I suggest that neither account can capture the role of the narrative order of the life of an individual or community plays in our evaluations of what happens at particular moments in time within those lives. In particular, the role of many environmental goods that are constitutive of community over generations is lost on the standard cost-benefit approach. Assumptions employed to compare costs and benefits in the commercial world are ill suited for evaluation of public projects over time. Their employment illustrates a more general problem concerning the disruption of narrative continuity between generations by commercial society.

Chapter 6 considers the central concept used to conceptualise relations between generations in recent environmental discourse, that of sustainability. The concept has been the object of some understandable scepticism, in particular where the adjective 'sustainable' is used to qualify nouns like 'growth' or 'development' given standard accounts of the nature of development. If the concept of sustainability is not to be a mere rhetorical flourish to documents that advocate standard models of economic growth it needs some more detailed specification. We need to be able to answer the questions – the sustainability of what, for whom and why? The dominant answers in the economic literature to those questions are reasonably straightforward: what is to be sustained is a certain level of human welfare understood in terms of preference satisfaction; current and future humans are those for whom it is to be sustained; to maximise human welfare over time or to meet the demands of intergenerational justice are the reasons why it is to be sustained. What is required in order that a certain level of human well-being is to be maintained or improved over time is that each generation pass on to the next generation a stock of capital that is as least as good as that it received from previous generations. The central argument then becomes one between two different accounts of what the requirement to maintain a level of capital involves, between proponents of 'weak sustainability' who hold that it is a matter of maintaining a total level of capital stocks, including both natural and human-made, and proponents of 'strong sustainability' who hold that it involves maintaining a critical level of specifically natural capital for which there are no human-made substitutes. In Chapter 6 I argue that if there are limits to the substitutability of natural and human-made capital, as defenders of strong sustainability claim, they need to be understood as part of a more general limits to substitutability that are not recognised in standard economic theory. There is an assumption of the ubiquitous substitutability of goods in mainstream economics that is a consequence of the theory of welfare it assumes. If one moves from a preference satisfaction account of well-being to more objectivist accounts that start from needs or human functional capabilities, then there are far more limits on the extent to which goods that meet different needs or capabilities can be substituted for each other. The chapter finishes by suggesting that the economic metaphor of 'capital' is ill suited to capturing the range of goods that are constitutive of the lives of individuals and community. In particular, they fail to capture the role of the narrative order in our appraisal of the quality of life. While the concept of sustainability may still have value in environmental discourse, it needs to be rescued from the market metaphors that have been employed to define the concept.

Part III looks at another of the central concepts of modern environmentalism, that of wilderness. Chapter 7 examines the way that the various uses of the concept highlight some of the complex and problematic relationships between environmentalism and the various processes of globalisation. The concept of wilderness has been central to deep green environmentalism in new world contexts. In the United States the defence of wilderness is often taken to define the environmentalist position, and a similar tendency is visible in places such as Australia. One central motivation of the appeal to wilderness in that context has been to protect places from the potential of unregulated commercial activity. The appeal to wilderness and wilderness designation of certain places is seen as necessary to their defence as sites free of commercial pursuits. Indeed the potential incursion is visible in the common experience in the United States of shopping malls and hotels running up to the boundaries of national parks where beyond those boundaries they cease and regulated movement begins. Conversely, for those in the 'Wise Use Movement' and 'Property Rights Movement' in the United States, the appeal to wilderness is likewise seen as a block to the proper expansion of market activity and a restriction on the exercise of property rights. That the course of the debate runs in this direction is a historically local phenomenon. In the sixteenth and seventeenth centuries the argument ran rather in the opposite direction. In the Lockean tradition, the designation of land as wilderness was rather appealed to as grounds for its appropriation of the land of the aboriginal populations. Since the land was uncultivated those populations had no claim upon them. They could be appropriated without complaint from those who lived there and brought into the world of commerce. The wilderness designation, far from being a block on the commercial appropriation of land, was a justification for that appropriation. The shift in the terms of the debate represents in part a shift of perception of lands and landscapes. However, it also points to the continuing problems with the idea of wilderness as a central concept of environmentalism. Historically the designation of land as wilderness disguised the pastoral activities of the populations whose land were appropriated. The lands were not wilderness but the homes of those who lived there before. At the same time, globally the use of the term wilderness continues to serve the appropriation of land of marginalised groups for national parks. Those parks themselves increasingly belong to the world of the global tourist industry. In this context there are some justifiable objections to the continued employment of the concept of wilderness by environmentalists. However, while the specific concept of wilderness has its problems in the defence of land from unrestricted commerce, this does not imply that either the need for that protection, nor many of the claims made in its defence fall. There is much that still survives in the environmentalist position when the particular problems with the wilderness concept are acknowledged.

The problems with the concept of wilderness do however raise another potential problem with modern environmentalism that has been at the centre of at least some liberal worries about some versions of green political thought. Green political theory, in so far as it moves beyond the specification of those primary environmental goods that are necessary condition for the pursuit of any conception of the good life, often appeals to a particular thick conception of the good life. Such appeals are often for example one source of criticisms of the patterns of consumption that drive modern market economies. Such patterns of consumption are not just taken to be unsustainable and hence to involve potential intergenerational injustice. They are often taken to be founded upon a mistaken view of what makes for a good human life and hence are to be discouraged for reasons independent of justice. It is not good for those who are the 'beneficiaries' of those forms of consumption. Similarly appeals to the value for humans of standing in a particular relation to nature undomesticated by human activity often appeal to an account of the human good. As the classical perfectionist liberal, J. S. Mill famously put it: 'It is not good for man to be kept perforce at all times in the presence of his species' (Mill, 1848, Book IV, ch. 6, section 2). For the modern anti-perfectionist liberal, the authoritarianism involved in the appeal to wilderness in modern conservation policy points to just what is wrong with appeals to a thick conception of the good in public affairs. Any view that sees the role of the public institutions to develop a particular conception of the good is incompatible with pluralism of modern societies. Whatever the defensibility of particular conceptions of the good in the private realm they have no role in the justification of policy in the public domain. Given the plurality of different conceptions of the good in the modern world, the consequences of the public pursuit of a particular contested conception of the good will be unjust and authoritarian. The justification of public policy has to restrict itself to a thin ethical vocabulary and a thin conception of the good.

Chapter 8 of the book responds to this position. On the one hand, the restriction of justification of public policy to a minimal thin language of valuation carries its own dangers of excluding voices from public deliberation. The dangers are apparent in the disputes between nature conservation bodies and marginalised communities discussed in Chapter 8 and more widely in the responses to economic valuation discussed in Part I of the book. In both what is apparent is a resistance in part to the restrictions in the languages of valuation that are employed in public decisions. Indeed a significant part of the problem of environmental justice concerns which languages of valuation can be included within public decision making and which are excluded. On the other hand there is no necessary incompatibility between pluralism and the employment of thicker languages of valuation in public deliberation. There is a widespread assumption that a thin moral language is a condition of conversation in a pluralistic society. Any thick language takes us to local and particular conceptions of the good. However, an examination of the examples of environmental disputes discussed in Chapter 7 suggests that this need not be the case, that thickening the language of valuation can be a condition of recognising shared goods specified in different cultural forms. Moreover, the plurality of goods themselves that are constitutive of a good life opens the space for a recognition of the plurality of the good life can be lived from within a thick account of human flourishing. Finally the role that narrative plays in our understanding of what it is for a life to go well itself is a condition of plurality. To make these points is not to deny the existence of conflict between different conceptions of the good, nor to deny that we need to find ways to resolve or live with such conflicts. It is to deny that invoking such conceptions in public decision making carries special dangers of an authoritarian politics or that a shift to a more minimal ethical language avoids such dangers.

The final part of the book considers some of the strengths and limitations of the deliberative alternatives to market-based approaches to environmental decision making. The environment has been one of the principal sites for the recent revival in deliberative models of democracy. For the deliberative theorist, democracy should not be understood as a market pursued by other means, as a mechanism for aggregating and efficiently meeting the given preferences of individuals. Rather it should be understood as a forum, a process through which preferences are formed and transformed through reasoned dialogue between free and equal citizens. As we noted in Part I of the book, the problem with market-based approaches to environmental decision making is that they are reason-blind. Environmental decisions, like other matters of public policy, are matters for the forum not the market. This developments of deliberative alternatives to economic models of democracy making have taken to two distinct forms. The first is in global civil society which has been given an increasingly deliberative reading and through the growth of regional and global social fora has taken overtly deliberative forms. The second is in the development of formal policy processes that are taken to be experiments in deliberative democracy and a response to the 'democratic deficit' of existing representative institutions: citizens' juries, consensus conferences, citizens' panels, focus groups and deliberative polls are typical examples. How far such putative processes of deliberation can in fact be said to be actually instantiate deliberative democratic procedures is however open to question.

The ideal of deliberative democracy is typically characterised in terms of dialogue between participants who are both formally and substantively equal. Rationally acceptable claims are those that can be redeemed in the absence of asymmetries of power, in conditions of equality in capacities to speak and be heard, in which the only force is that of the better argument. Clearly actual experiments in deliberation can and do depart markedly from such ideal conditions. Within deliberation capacities to speak and be heard differ across class, gender and ethnicity. Moreover, recent formal processes are often exercises in controlled conversation. The identity of participants, the opening and closure of the processes, the agenda for deliberation and the institutional afterlife of the results of deliberation are not controlled by participants. This need not entail a lack of legitimacy for the results of deliberation but it does allow that the processes can be used strategically by powerful actors, for example to close down wider deliberation or to capture and domesticate oppositional voices. Formal deliberative procedures can and are used by powerful actors to legitimise actions, and indeed can do so more effectively than expert-based procedures, such as cost-benefit analysis, which lack democratic legitimacy. At the same time, the informal deliberative processes of civil society have their own asymmetries of power and are themselves also open to subversion by powerful actors.

To make these points is not to criticise the deliberative model of democracy as such. Indeed the deliberative theorist can properly respond that these are internal criticisms of the practice of deliberation from within a deliberative framework. Deliberative institutions fail from within that perspective to the extent to which their procedures are distorted by asymmetries of power and strategic action. Deliberative theory provides the normative basis from which an immanent criticism of the failings of putatively deliberative institutions can proceed. There is much that is right with this deliberative response to the failings of actually existing deliberation. However, those failings also raise difficulties for the standard approaches to deliberative democracy that have developed in different ways from the work of Habermas and Rawls. The final three chapters of the book consider three of these sets of problems around such approaches: first, problems concerning the forms of communication that such approaches include and exclude; second, problems around who can be present in deliberation and the ways that those are absent are represented; third, problems about the degree to which different forms of inequality in voice can be addressed by the predominantly political and cultural focus of liberal accounts of deliberation.

Deliberative criticisms of actually existing deliberation presuppose a particular set of criteria of good deliberation. The work of Habermas and Rawls which is the starting point for much recent deliberative theory starts from common Kantian perspective on the conditions deliberation ought to ideally meet. They inherit an ideal of the free public use of reason which Kant takes to define the enlightenment project. The broad contours of that enlightenment heritage I share against its recent critics. What is more questionable however is the particular intellectualist account of the public use of reason that also has Kantian roots. There is a widespread criticism of Habermasian and Rawlsian account of deliberation for excluding from legitimate deliberation a range of forms of communication and consequently also a range of voices in public deliberation. To the extent that those criticisms can be justified, it is in virtue of the inadequacies of the intellectualist account of deliberation that both inherit from Kant. In particular, two features of public argument which a more Aristotelian account renders central and legitimate parts of deliberative processes are dealt with badly by the Kantian tradition: the role of testimony and the appeal to the emotions. Both for Kant are inconsistent with the maturity and autonomy required for the ideal agents of the enlightenment. However, there are good grounds for thinking that neither can nor ought to be eliminated from public deliberation. The role of scientific expertise in the assessment of risks means that in the environmental sphere most ordinary citizens need to rely on the testimony of others. The central question for public deliberation is not the direct assessment of evidence and argument, the assessment of whose testimony is credible and trustworthy – of deciding who to believe in what institutional conditions. Likewise, the appeal to the emotions to motivate public action is central to the environmental sphere as it is to others spheres of political life. In Chapter 10 I argue that an Aristotelian account of public deliberation gives us a better understanding of the conditions in which such appeals to testimony and to the emotions are legitimate. Moreover, while this Aristotelian account of deliberation offers a less heady and more socialised account of the capacities and powers of the mature ethical agent, it is consistent with the central enlightenment values that the Kantian theorists aim to defend.

Chapter 11 turns to problems of representation. One dimension of deliberative criticisms of market-based approaches to environmental decision making has to do with failings of representation: the poor are under-represented through willingness to pay measures; future generations and non-humans are represented at best precariously through the preferences of current consumers. However, deliberative approaches have their own problems. Marginalised groups are often under-represented both in formal deliberative fora and the informal processes of civil society. Those chosen to speak within formal deliberative institutions are not normally accountable to those they are often taken to represent. Asymmetries of power and voice exist between the different associations of civil society and movements are not always accountable to the marginalised groups on whose behalf they speak. The representation of future generations and non-humans raises still less tractable difficulties for deliberative theories of democracy. Their direct voice is necessarily absent. Hence in deliberative institutions their voice is at best indirectly represented by those who claim to speak on their behalf without accountability to them. The necessary absence of direct representation raises well-rehearsed major problems concerning the ethical and political legitimacy of decisions made in the absence of their voice. It also raises questions about the legitimacy of the claims made by environmental organisations and activists to speak on their behalf. Finally the quality of deliberation in formal and informal contexts is often in tension with the quality of deliberation: the quality of deliberation improves in smaller fora, the quality of representativeness in larger fora and the two demands pull in the opposite direction. The problem of representativeness is often misstated in the context of the social sciences – it is not a problem of statistical representativeness, but rather political legitimacy. Whose voice is present, under what description and under what conditions of legitimation?

Chapter 11 outlines the main contours of answers to those questions. The central claim that can be made for deliberative as against market-based representation is that it must survive Kant's publicness condition. Reasons offered in a deliberative assembly, unlike the private preferences expressed in market acts, must be able to survive being made public. The publicness condition forces participants to offer reasons that can withstand public justification and hence to appeal to general rather particular private interests. Hence reasons for action that appeal to wider constituencies of interest – including those of future generations and non-humans – are more likely to survive in public deliberation than they are in private market-based methods for expressing preferences. Thus goes the claim for deliberation. However, the claim does not resolve the question of the grounds individuals can claim to speak for others who cannot speak for themselves, who can authorise no one to speak for them and to whom no accountability is possible. The legitimacy of representation in such contexts is reduced to epistemic claims. Those who claim to speak on behalf of those without voice do so by appeal to their having knowledge of their objective interests and special relations of care for them. However, the claims made are clearly open to dispute. The disputes are of particular significance where the representatives of nature have too much voice, rather than too little. Hence for example the disputes between nature conservation bodies and marginalised communities discussed in chapters 7 and 8. To raise those conflicts is not to suggest how they might be resolved. Indeed, against the assumption to be found in at least some of the deliberative literature, there maybe no resolution, no set of judgements on which deliberation would converge in ideal conditions. Different goods of human life conflict and there is not always a resolution to be had. Consensus in actual debate is often a sign not of reasoned deliberation but silenced voices. The virtue of publicness in deliberation may not be that it is a condition of convergence of judgements, but rather that it is a condition of conflicts and dissensus can be made visible.

Chapter 12 considers issues of power and voice in more detail. The legitimacy of deliberative procedures depends upon a degree of formal and substantive equality among participants. How is such equality to be realised? The standard move within liberal theory is to attempt to protect the political domain of deliberation from wider structural inequalities in the distributions of power and resources: deliberative institutions are to be autonomous from those wider patterns of social and economic power. One major thrust of recent criticism of liberal deliberative theory has been over its failure to recognise the ways that different forms of communication are excluded and different groups fail to be recognised within liberal models of deliberation. However, those criticisms themselves have often tended to stay at a purely discursive level. The result is a purely symbolic or cultural politics which fails to address the ways in which the structural imperatives of markets place constraints on the actual decisions of actors. One reason for that shift has been the loss of confidence in alternatives to liberal market economies given the undoubted failures of central planned economies. One possible response has been to attempt to take to deliberation into the economy in combination with more associational models of socialism. That response is however open to a version of Hayek's epistemic objections to centrally planned economies. If, as Hayek claims, there are forms of tacit or practical knowledge that cannot be passed on to a central planning system because they are not open to articulation in propositional form, then neither will they be open to articulation in a deliberative setting. Hence, models of decision making that rely purely on deliberative procedures will be open to the some of the same objections as are made in the epistemic case against planning. The argument has been employed more widely against deliberative approaches to environmental decision making. Indeed Hayek's own epistemic arguments were aimed not only against planned economies but also against the tradition of thought that issued in ecological economics. In Chapter 12 I respond to these arguments. Their force in part is based on the overly intellectualist model of deliberation itself that we criticised in Chapter 10. If one rejects that model one can recognise the ways in which deliberative institutions themselves call upon habitual forms of behaviour and practical knowledge that is not itself open to articulation in propositional form. In defending a more associational and deliberative model of economic life, I will show that the particular arguments about the limits of market solutions to environmental problems, discussed in the first part of the book, are relevant to the wider current debates not just about the institutional and structural conditions for an environmentally rational society, but also for alternatives to globalised market relations discussed at the outset of this introduction.

Part I

Environmental goods and the limits of the market

1 Markets and the environment

The solution is the problem

1.1 The solution is the problem

What is the source of our environmental problems? Why is there in modern societies a persistent tendency to environmental damage? From within the neoclassical economic theory there is a straightforward answer to those questions: it is because environmental goods and harms are unpriced. They come free. Indeed, Arrow (1984) claims this to be a thesis peculiar to neoclassical economics: 'The explanation of environmental problems as due to the nonexistence of markets is . . . an insight of purely neoclassical origin' (Arrow, 1984, p. 155). Given that the source of environmental damage is that preferences for environmental goods are not revealed in market prices, then the solution is to ensure that they are. Hence the claim runs that we should either bring environmental goods into actual markets through an extension of tradable property rights to environmental goods, or alternatively we should construct shadow prices for environmental goods by ascertaining what individuals would pay for them were there a market. The construction of shadow prices is carried out either indirectly – by inferring from some proxy good in the market such as property values an estimated price for environmental goods or by using the costs incurred by individuals to use an environmental amenity to estimate values – or directly by the use of contingent valuation in which monetary values are estimated by asking individuals how much they would be willing to pay for a good or accept in compensation for its loss in a hypothetical market. Those economic values can then be entered into a cost-benefit analysis for any proposed projects such that their full benefits and costs can be ascertained. The standard Kaldor-Hicks test of efficiency employed in cost-benefit analysis can then be applied. A proposed project constitutes a welfare improvement if the gains are greater than the losses, so that the gainers could compensate the losers and still be better off. The optimal decision is that which produces the greatest benefits over costs. It is through extending prices to environmental goods so that their 'true' value can be discovered that the road to resolving environmental problems lies. Thus goes the neoclassical position.

That position runs up against a view which runs in entirely the opposite direction, that our environmental problems have their source not in a failure to apply market norms rigorously enough, but in the very spread of market mechanisms and norms.

The source of environmental problems lies in part in the spread of markets both in real geographical terms across the globe and through the introduction of market mechanisms and norms into spheres of life that previously have been protected from markets. That view has its own long history. One version is to be found in the eighteenth-century civic humanist criticism of rising commercial society which centred on the claim that the commercial mobilisation of land would undermine links between generations. A distinct and more recent version is defended by Karl Polanyi in his arguments around the environmental effects of the mobilisation of land and labour as commodities in the marketplace. Withdrawing both from the constraints of ethical and social norms would result in the destruction of the environment and the social dislocation of humans:

> To allow the market mechanism to be sole director of the fate of human beings and their natural environment . . . would result in the demolition of society. For the alleged commodity 'labor power' cannot be shoved about, used indiscriminately, or even left unused, without affecting also the human individual who happens to be the bearer of this peculiar commodity. In disposing of man's labor power the system would, incidentally, dispose of the physical, psychological, and moral entity 'man' attached to that tag. Robbed of the protective covering of cultural institutions, human beings would perish from the effects of social exposure; they would die as the victims of acute social dislocation through vice, perversion, crime and starvation. Nature would be reduced to its elements, neighbourhoods and landscapes defiled, rivers polluted . . . the power to produce food and raw materials destroyed.
>
> (Polanyi, 1957a, p. 73)

Given a version of this second view, the neoclassical project of attempting to cost all environmental goods in monetary terms becomes an instance of a larger expansion of market boundaries. The proper response is to resist that expansion, be this in the spirit of resistance to market society or more modestly to maintain the proper boundaries between spheres.[1] The market can expand its boundaries in at least two different ways (Keat, 1993, pp. 13ff.). First, items that are considered inappropriate for sale might become directly articles for sale on the market: consider the sale of bodily parts, sexual services, reproductive capacities, votes, political office, the means of salvation and so on. Second, relations, attitudes, forms of evaluation and the like typical of the market might be transferred to other spheres: for example, while university education in the United Kingdom is not yet a commodity that is directly bought and sold on a free market, there has been an increasing replication of the language and relations of the market in for example the treatment of students as 'consumers' and the reconceptualisation of the relation of teacher and student in contractual terms. The neoclassical response to environmental problems raises issues on both fronts. Free market solutions to environmental problems, that attempt to extend tradable property rights to environmental goods raise questions of market boundaries of the first kind. The practice of economic valuation for the purpose of cost-benefit analysis raises concerns about market expansion in its second form: it represents an incursion of market-based norms and modes of arriving at choices into spheres where they are inappropriate. In the following chapters I will criticise both neoclassical responses.

In this chapter I want to consider some of the problems with the neoclassical project which we will deal with in this section of the book by reflecting on examples of resistance to demands to put a price on environmental goods, be this actual or hypothetical. Consider a protest to an actual request to price the environment which captures a number of thoughts to which I will return throughout this book. How much would you be willing to accept in compensation for the loss of your homeland by a dam project? Here is an excerpt from a letter from an inhabitant of the Narmada Valley in western India, threatened with displacement as a result of the Sardar Sarovar Dam, written to the Chief Minister of the state government.

> You tell us to take compensation. What is the state compensating us for? For our land, for our fields, for the trees along our fields. But we don't live only by this. Are you going to compensate us for our forest? . . . Or are you going to compensate us for our great river – for her fish, her water, for vegetables that grow along her banks, for the joy of living beside her? What is the price of this? . . . How are you compensating us for fields either – we didn't buy this land; our forefathers cleared it and settled here. What price this land? Our gods, the support of those who are our kin – what price do you have for these? Our adivasi (tribal) life – what price do you put on it?[2]

The response is I think a quite proper one. It illustrates general problems with the very attempt to extend prices to environmental goods and more particular problems with the use of any compensation test in appraising projects.

1.2 The limits of monetary valuation

Running through the protest to the compensation offered in the letter above is a resistance to the very acting of pricing certain goods. What are the problems with the project of extending prices that the passage raises? One objection to the project which is central to this passage is that there are certain social relations and evaluative commitments that are constituted by a refusal to put a price on them – there exist what are sometimes called 'constitutive incommensurabilities'.[3] In particular, our relations to the intergenerational communities to which a person belongs, for example to kin who have gone before us and will follow after, cannot be expressed in monetary terms. This objection I will discuss in more detail in section 1.3 and examine in detail in Part II of the book. Second, there are objections to the treatment of the environment as an alienable property, to the extension to the places that matter to individuals and communities of a particular set of liberal property rights. I discuss this briefly in section 1.4 and in more detail in Chapter 3. Third, there are issues of equity. These are not directly raised in this quotation, but which were central to the use of monetary prices in the case of the Sardar Sarovar Dam and the levels of compensation offered. Given a monetary valuation, it is usually the case that 'the poor sell cheap'. In the case in question, the cost of displacement is calculated on the basis of income forgone: given low income the costs are low. I discuss these issues of equity briefly in section 1.4 and in more detail in Chapter 3. Finally, the very appeal to the reasons why this particular environment matters to those who live there points to a general problem with the use of monetary measures, that is that they are reason-blind. The reasons for the valuation of environmental goods articulated here are rendered invisible in market prices. Willingness to pay at best picks up the intensity of preferences of individuals. It does not pick up the soundness of reasons for preferences and the degree to which they should weigh with a wider community. The point is at the heart of deliberative criticisms of monetary valuation. I discuss this in sections 1.5 and 1.6 and I will return to it at length in the final part of the book.

1.3 Constitutive incommensurabilities

A useful starting point to the issue of constitutive incommensurability is that of another older monetary valuation study of a non-environmental good. I do so in the spirit of historical scholarship. I offer what must be one of the earliest reports of such contingent valuation surveys. The report comes from Herodotus' histories:

> When Darius was king of the Persian empire, he summoned the Greeks who were at his court and asked them how much money it would take for them to eat the corpses of their fathers. They responded they would not do it for any price. Afterwards, Darius summoned some Indians called Kallatiai who do eat their parents and asked in the presence of the Greeks . . . for what price they would agree to cremate their dead fathers. They cried out loudly and told him to keep still.
>
> (Herodotus, *Histories*, 3.38, translation from McKirahan, 1994, p. 391)

I start by noting an obvious contrast between Darius' survey of individuals' willingness to accept a price on a good and the willingness to pay or accept surveys of his modern economic counterpart. Darius' survey aims to elicit protest bids. The story would have been somewhat ruined if the Kallatiai had responded by putting in a realistic price. The reason why Darius elicits the protests is to reveal the commitments of the individuals involved. One exhibits commitment to some good, here one's dead kin, by refusing to place a price upon it. In contrast, the modern economist begins by ignoring all protest bids: these together with strategic responses are laundered out of the responses to leave us with just those relevant to a calculation of the welfare benefits and costs of the project. Part of the problem here is with the view of monetary prices that economists in both the neoclassical and Austrian traditions assume. Monetary transactions are treated as exercises in the use of 'a measuring rod', that of money. That view is false. Monetary transactions are social acts which have a social meaning.

Certain kinds of social relation and evaluative commitments are constituted by particular kinds of shared understanding which are such that they are incompatible with market relations. Social loyalties, for example, to friends and to family, are constituted by a refusal to treat them as commodities that can be bought or sold. Given what love and friendship are, and given what market exchanges are, one cannot buy love or friendship. To believe one could would be to misunderstand those very relationships. To accept a price is an act of betrayal, to offer a price is an act of bribery (Raz, 1986, pp. 345ff.; O'Neill, 1993a, pp. 118–122). Similarly ethical value commitments are also characterised by a refusal to trade. Hence the response to the offer of compensation in the passage on the Sardar Sarovar Dam. The very acceptance of a price on relations of kinship and a way of life is to betray them, the offer to corrupt. The use of the term 'compensation' in this context disguises what is going on. Central to the criticism of the spread of market relations into other social spheres is the corruption this entails of social relations that are central to a good human life. The social meanings that are constitutive of markets are such that they are incompatible with other relationships – and for this reason what matters is not just whether an object is sold but the spread to other spheres of the forms of understanding that are constitutive of markets.

For this reason, it is unsurprising to find similar responses to the demand for contingent valuation studies. Consider for example the following responses expressed in in-depth discussion groups conducted by Burgess et al. (1995) among groups who had been asked contingent valuation questions about a Wildlife Enhancement Scheme (WES) on Pevensey Levels in East Sussex undertaken on behalf of English Nature.[4] While the case is much less pressing than the Sardar Sarovar Dam in terms of its social impacts, it still has a dimension that concerns social relationships, in particular the communities we belong to over time. The environment matters as a place that embodies particular relations of the past and through which relations to the future are expressed. Hence some of the responses to the Pevensey Levels survey that emerged in the in-depth group discussions of the local residents. The value of the Levels lies in part in the very particular history of social relations it embodies and its destruction matters in virtue of disturbing physically embodied memories. The comment of one of the residents, Kate, captures a fairly common reaction to the destruction of local landscapes and habitats:

> It's recalling memories of my childhood down on the beach at the Crumbles, we used to spend a lot of time down there with my dad, sea fishing, and it was all the plants and stuff he was just mentioning. I'm thinking now when I drive through there in the morning that there's *none left at all*. It's quite sad that these things have gone.
>
> (Burgess et al., 1995, p. 34, emphasis in the original)

Correspondingly, the environment is expressive of social relations between generations. It embodies in particular places our relation to the past and future of communities to which we belong.[5] And it is that which in part activates the protests to the demand to express concern for nature in monetary terms, including protests from some who may have actually responded 'legitimately' to the survey. Typical is the following response of one respondent, who didn't actually put in a protest bid:

> it's a totally disgusting idea, putting a price on nature. You can't put a price on the environment. You can't put a price on what you're going to leave for your children's children . . . It's a heritage. It's not an open cattle market.
>
> (Burgess et al. 1995, p. 44)

The point here is that an environment matters because it expresses a particular set of relations to one's children that would be betrayed if a price were accepted upon it. The treatment of the natural world is expressive of one's attitude to those who will follow you.

The problem here concerns the attempt to express values in terms of money prices. However, while the use of money values raises particular problems, the problem lies not just in the use of the money measure, but in a more general assumption that underpins cost-benefit analysis, i.e. that good decisions require a common measure to trade off gains and losses so that choice becomes an exercise in mathematics. Cost-benefit analysis assumes that rational choice requires a single measure of value and that money prices just happen to be the one.

> Physical accounts *are* useful in answering ecological questions of interest and in linking environment to economy . . . However, physical accounts are limited because they lack a common unit of measurement and it is not possible to gauge their importance relative to each other and non-environmental goods and services.
>
> (Pearce et al., 1989, p. 115, emphasis in the original)

The general assumption that rational choice requires commensurability is false. The environment is a site of conflict between competing values and interests, and institutions and communities that articulate those values and interests. These cannot be reduced to a single measure, whether monetary or otherwise.[6] Moreover, there is no reason to assume that rational choices demand commensurability (O'Neill, 1997a, 1998a ch. 9; Martinez-Alier et al., 1998, 1999, 2001). They demand debate and argument and not sums to resolve them. I will return to this more general point in Chapter 2.

1.4 Property rights and equity

Protests to demands to price environmental goods are often articulated in terms of a rejection of the assumptions about property rights that the demand for pricing involves. There are at least three issues here. The first concerns the conception of property rights that economic valuation assumes. Economic valuation either hypothetical or actual requires that environmental goods be understood as commodities which could in principle be exchanged in markets. As such valuation presupposes that environmental goods can be brought under a liberal conception of property or full property rights required for market exchange. The development of market economies requires property rights that involve exclusivity and alienability. Protests to valuation studies often reject the questions because they do not think that environmental goods should be treated as alienable goods. The point is implicit in the protest concerning the Sardar Sarovar Dam with which I started. 'How are you compensating us for fields either – we didn't buy this land; our forefathers cleared it and settled here. What price this land?' The point here concerns issues about the acquisition and subsequent alienability of goods. The land wasn't something bought in markets and hence it isn't something that can be alienated in subsequent exchange. It is not an object of liberal ownership.

A similar point runs through many of the protests recorded in the Pevensey Levels discussion. When it comes to the environment in relations between generations, the issue of price is rejected because as one resident puts it, 'it's not ours to sell' (Burgess et al., 1995, p. 44). The point here is that issues of buying and selling arise only if one assumes one has rights to alienate the goods in question. If one understands anyone's relations to environmental goods over time to be those of a person with use rights but not rights to alienate the good, then the question of willingness either to pay or to accept do not arise in the first place. If the environment belongs to your children it's not yours to buy or sell. The point here is I take it an ethical one, not a legal one. Clearly, one can in fact buy and sell property – the point is that to sell and leave your children with nothing is to betray their claims on a good that one has within one's power. As a resident in Pevensey Levels study remarks, it's 'not just for you. It's your children and your children, their heritage' (Burgess et al., 1995, p. 44). Given that the goods in question are such that only use rights are morally legitimate, the question of payment does not arise.

The second point that the Sardar Sarovar Dam letter highlights concerns the nature of the good. As Vatn and Bromley (1994) note: 'A precise valuation demands a precisely demarcated object. The essence of commodities is that conceptual and definitional boundaries can be drawn around them and property rights can then be attached – or imagined' (Vatn and Bromley, 1994, p. 137). With environmental goods, it is often not possible to demarcate them in that way and treat them as something to which discrete property rights could be assigned. What is the good to which rights are being attached – the fields, the land, the trees along the fields, the forest, the river, relations to kin, a way of life?

The third assumption that is often questioned in protests to economic valuation exercises is just who has rights. An economic valuation exercise assumes a structure of rights over the goods in question and it is sometimes this very structure of rights that is the point at issue. The existence of such conflicting rights claims has effects on willingness to pay responses, for an answer to it is required to ascertain who is supposed to pay and who is to receive payment for the maintenance of a particular good. But the fact that economic tools such as cost-benefit analysis presuppose the de facto distributions of property entitlements and income also raises well-rehearsed problems of equity.

The distributional problems associated with the employment of either actual or hypothetical market prices to arrive at decisions concerning different projects are well known. To employ unmodified willingness to pay measures entails that the poorer you are the less your preferences count. More generally, if you rely on market mechanism, the poor sell cheap. For example, if in a contingent valuation a person who is unemployed answers honestly and non-strategically that she cannot afford anything for a particular good then she is effectively silenced no matter how much she cares for the good in question.[7] That person's care cannot count as much as those who can afford to express their care in additional monetary payments for environmental goods. Or in the case of mass displacements to establish dams, if those moved are poor the 'human cost' is little. In the notorious cost-benefit analysis used to justify the Narmada Valley project, the 'cost' to the displaced was computed at two years' family income multiplied by a factor of 1.5 (Alveres and Billorey, 1988, pp. 19–20). Now there may be possible theoretical responses to this. One response from within neoclassical welfare economics is to readjust the responses by giving differential weight to the preferences of the poorer on the grounds that the marginal utility of money is greater for the poor than it is for the rich. The social welfare is defined as the sum of net benefits to each party thus adjusted. It is not clear how far this addresses the problem of respondents who fail to register at all their preference for a good. You can't give additional weight to a zero bid. In the actual practice of cost-benefit analysis such adjustments are rarely employed. Distributing burdens to the poor entails lower official costs. I will examine some of these problems further in Chapter 3 and some general issues of equity and the environment in chapters 7 and 8.

1.5 Reason blindness

Economic valuation and the process of cost-benefit analysis it informs are reason-blind. It is preferences that count, not the reasons why the preferences are held. No willingness to pay or accept a figure captures the considerations that are offered for some 'preference', say about some particular wetland, or more whether or not a community should be moved from its home to make way for a dam. The kinds of consideration about the value of a river that are offered in the passage from local inhabitant living by the Narmada disappear from view in an exercise in monetary valuation. Monetary valuations at best pick up the strength and weakness of the *intensity* of a preference. The strength and weakness of the *reasons* for a preference are not registered. Preferences grounded in ethical, aesthetic, scientific judgement, judgements about a place and its role and history in the life of a community are treated as on par with preferences for this or that flavour of ice cream. Judgements are treated as expressions of mere taste to be priced and weighed one with the other. They do not have to undergo the test of being able to survive public deliberation. Consequently the process offers conflict resolution and policy without rational assessment and debate.[8]

The arguments about the place of cost-benefit analysis in public decision making are in part arguments about how far public life itself should be conducted according to the norms of choice in market transactions. It is an argument about the nature of rational decisions and of the boundaries between markets and politics.

Cost-benefit analysis inherits its reason blindness of the market itself. As Hirsch memorably puts it, the market is 'in principle unprincipled': 'In the modern liberal view, the socio-economic system is seen as amoral' (Hirsch, 1977, p. 119). The market is not just amoral but arational at the level of preferences. In market transactions, the reasons for preferences are irrelevant. Indeed, many who defend market economies assume that preferences and values cannot answer to reason at all. They are mere expressions of taste. Economists in both the neoclassical and Austrian traditions typically assume a non-cognitivist view of values. Ends are treated as wants, and no judgement of their inferiority or superiority is allowed to enter criteria of choice. Markets offer decisions without dialogue. Coordination in markets is achieved by 'exit' – some goods find a market, others do not. Voice is not required (Hirschman, 1970).

Rationality on this account enters only in an instrumental form, as a set of procedures for effectively meeting ends that are given. Ideally it is algorithmic. On the algorithmic conception of practical rationality for a decision-making process to be rational it must be the case that there exists (i) a set of technical rules which are such that (ii) when given a suitable description of a different object or state of affairs they yield (iii) by a mechanical procedure (iv) a unique and determinate optimal outcome. The great virtue of the market on this view is that through the universal unit of money it offers an algorithm for calculating the ratio of benefits to costs. The point is stated with clarity by Schumpeter (1987):

> [C]apitalism develops rationality and adds a new edge to it . . . [I]t exalts the monetary unit – not itself a creation of capitalism – into a unit of account. That is to say, capitalist practice turns the unit of money into a tool of rational cost-profit calculations, of which the towering monument is the double-entry bookkeeping . . . And thus defined and quantified for the economic sector, this type of logic or attitude or method then starts upon its conqueror's career subjugating – rationalizing – man's tools and philosophies, his medical practice, his picture of the cosmos, his outlook on life, everything in fact including his concepts of beauty and justice and his spiritual ambitions.
>
> (Schumpeter, 1987, pp. 123–124)

Correspondingly, there is a widespread assumption in both neoclassical and Austrian traditions of economics that where a unit of calculation is absent, rational choice is not possible. Rational choice consists in 'trading off' costs and benefits through the use of a common unit of comparison. Rationality requires commensurability. Indeed, it is often argued that it requires monetary commensurability.

This assumption that rational choice in the market offers norms for all decisions is that which the arguments about the role of cost-benefit analysis in environmental put into question. For grounds that I've outlined here, there is no reason to assume that the use of cost-benefit calculations based on monetary valuations will issue in rational choice. There is no reason to assume that the procedures of rational choice require commensurability, the existence of a common unit of measurement through which 'costs' and 'benefits' can be traded off against each other.

What then are the features of rational decision making in the absence of commensurability and algorithms? Three are I think worth emphasising here. The first is that there is an ineliminable role of judgement. The power of judgement is required in the application of any universal rule to particular cases and cannot itself under pain of an infinite regress be understood as the application of a rule.[9] Second, procedural norms of practical reason apply. Procedural accounts of practical reason take an action to be rational if it is an outcome of rational procedures: 'Behaviour is procedurally rational when it is the outcome of appropriate deliberation' (Simon, 1979, p. 68). Raz's account of rational action under conditions of incommensurability can be understood as procedural in this general sense: 'Rational action is action for (what the agent takes to be) an undefeated reason. It is not necessarily action for a reason which defeats all others' (Raz, 1986, p. 339). These norms of rationality are implicit in the criticisms I have made about the reason-blindness of cost-benefit analysis. Rational behaviour is that which emerges from deliberation that meets the norms of rational discussion. Given a procedural account of rationality, what matters is the development of deliberative institutions that allow citizens to form preferences through reasoned dialogue, not institutions for aggregating given preferences to arrive at an 'optimal' outcome. Third, behaviour ought to meet norms of expressive rationality. Expressive accounts characterise actions as rational where they satisfactorily express rational evaluations of objects and persons: 'Practical reason demands that one's actions adequately express one's rational attitudes towards the people and things one cares about' (Anderson, 1993, p. 18). Actions are not just instrumental means to an end, but also a way of expressing attitudes to people and things. The point underlies the rationality of constitutive incommensurabilities noted earlier. If I care about something, then one way of expressing that care is by refusing to put a price on it.[10]

1.6 Two models of democracy

The argument between algorithmic and deliberative conceptions of practical reason in public life is an argument about the nature of public life itself. And that argument is in part an argument about the boundaries of markets. How far should political life be governed by market norms? Defenders of cost-benefit analysis often appeal to a market view of democracy. Democracy, like the market, is a procedure for aggregating and effectively meeting the given preferences of individuals. Prices and votes on this view are different versions of the same thing. Indeed, contingent valuation has been presented a real instantiation of qualitative democracy, that is that form of democracy in which citizens are able to register not just support for a particular position, but also the degree to which they care about it. Through contingent valuation individuals are able not simply to 'vote' for a good by expressing support for it, but also to express through their response to a willingness to pay question their degree of concern. The argument is put with characteristic clarity by Pearce et al. in the *Blueprint for a Green Economy*:

> [T]he attraction of placing money values on these preferences is that they measure the *degree* of concern. The way in which this is done is using, as the

> means of 'monetization', the willingness of individuals to pay for the environment. At its simplest, what we seek is the expression of how much people are willing to pay to preserve or improve the environment. Such measures automatically express not just the fact of a preference for the environment, but also the intensity of that preference. Instead of 'one man one vote', then, monetization quite explicitly reflects the depth of feeling contained in each vote.
>
> (Pearce et al., 1989, p. 55, emphasis in the original)

On this view there is no essential difference between markets and politics, prices and votes. Indeed, politics aims at ideal market outcomes by other means, and cost-benefit analysis offers one technical tool to realise that aim.

This model of market model of democracy raises the problems of both equity and reason. It raises the problem of equity – votes are distributed according to income. It raises problems of rationality. It is against this market-based conception of political life that recent revival of deliberative democracy has been aimed. Against the picture of democracy as a procedure for aggregating and effectively meeting the given preferences of individuals, the deliberative theorist offers a model of democracy as a forum through which judgements and preferences are formed and altered through reasoned dialogue between free and equal citizens. From this perspective the reason-blindness of cost-benefit analysis is a source of its problems. The features of practical rationality I outlined above, the ineliminable role of judgement and the appeal to procedural and expressive norms of rationality could all be taken to underpin the employment of more deliberative institutions for environmental choices.

To defend a deliberative conception of public life is not to say that current political institutions are instantiations of deliberative democracy, still less that recent experiments in deliberative democracy successfully realise deliberative procedures. For reasons that I develop at length in Part IV of the book there are reasons for scepticism. A feature of modern political life is the degree to which it is being itself invaded by the norms and culture of the market: one does market research to form and sell policies, not to persuade by argument. Sound-bites are not arguments, even with suppressed premises. They conform rather to Lewis Carroll's snarxist theory of truth: 'what I tell you three times is true' (Carroll, 1962, p. 38). For a sound-bite you need to say it a few more times than that, but the principle is the same. Moreover, some institutions that are presented as experiments in deliberative democracy, such as focus groups, properly belong to this marketisation of politics rather than a deliberative alternative. The origin of the focus group technique in market research is not without its implications. It is often employed in political practice not to allow deliberation to take place but rather to close it down. It is employed to gather information of likely responses to different potential policies and actions not to open up debate, but to anticipate and forestall it. It belongs to the marketisation of politics. So also does cost-benefit analysis: what matters is the intensity of the preferences, not the soundness of the reasons, and preferences count only to the extent that they allow of monetary expression. Given the drift in modern politics towards an audit culture in which that which comes without a money figure disappears from the political processes, there is an understandable temptation for the environmentalist to be pragmatic about this and follow the trend to marketisation on the grounds that it gets results. This is what the environmental economist offers. The temptation should however be resisted. This is, in part, because the quality of public life matters as well as the quality of the environment. However, in the long term the quality of the environment itself is likely to depend on that of public life.

1.7 The limits of compensation

I want to finish by considering some more particular problems with the very idea of a 'compensation test' in cost-benefit analysis that the letter on the Narmada Valley highlights. Standard cost-benefit analysis employed in project evaluation appeals to the Kaldor-Hicks compensation test. The test is introduced to avoid the problems of applying standard Pareto efficiency tests in contexts where there are both winners and losers. On the standard Pareto criterion a proposed situation A represents an improvement over another situation B if someone prefers A to B but no one prefers B to A. Where there are both winners and losers the criterion does not apply. Hence the shift to potential Pareto improvements employed in the Kaldor-Hicks test: a situation A is an improvement over B if the gains are greater than the losses, so that the gainers could compensate the losers and still be better off. The compensation test is taken to avoid the difficulties in appealing to actual Pareto improvements, while retaining the claimed advantage of the Pareto criterion, that it allows us to make judgements about welfare improvements without making interpersonal comparisons of welfare.[11] The compensation test is open to a number of well-known technical objections. However, at a more basic level it is open to the obvious problem that hypothetical compensations are no compensations at all. As Sen (1987) pithily puts the point:

> The compensation principle is either redundant – if the compensation is actually paid then there is a real Pareto improvement and hence no need for the test – or unjustified – it is no consolation to losers, who might include the worst off members of society, to be told that it would be possible to compensate them even though there is no actual intention to do so.
>
> (Sen, 1987, p. 33)

Clearly one response to this objection that hypothetical compensations are no compensations is to move back to actual compensations. Protests to the very request to accept compensation point to difficulties with that move. There are goods for which no monetary compensation is acceptable. Neither is that protest irrational. For some goods there are no substitutes.

The supposition that compensation is widely possible is based upon a particular model of welfare that underpins neoclassical economics.[12] Welfare consists in the satisfaction of preferences. Goods are substitutable for each other as long as total preference satisfaction levels remain unchanged. The textbooks assume a wide degree of substitutability. A welfare loss in one dimension of goods can be compensated for in a gain in another provided the overall level of preference satisfaction is sustained. It is this assumption that underpins the indifference curves of neoclassical economics – for any incremental loss a particular good that satisfies one end there is an incremental gain in another that compensates for that loss and maintains preference satisfaction. The agent is indifferent between the two bundles of goods. Hence, goods can be traded off against each other at the margin. For a loss on one dimension of valued goods – say of a place – there are gains in other dimensions which can compensate for that loss. Given this account there will be a sum of money that can be paid for a good such that a marginal welfare loss on one dimension of goods can be compensated for by gain in others so that total welfare remains unchanged or better still improved. On that view the everyday claim, 'nothing can compensate me for this loss', is at best an exaggeration. Properly understood, there will normally be a particular level of compensation that will return an individual to the level of welfare they enjoyed before the loss. This account assumes a preference satisfaction account of welfare.

There are good reasons for rejecting this preference satisfaction account of welfare. A crude preference satisfaction account of welfare that identifies well-being with the satisfaction of whatever preferences people happen to have is open to obvious objections. It does not allow for the possibility of making a mistake about what is for our own good – and clearly we only too often do make such mistakes. There are many things we desire which when fully informed about the object we would no longer value or desire. We desire a particular food on the basis of the belief that it is healthy and tasty, and it turns out in fact to be a carcinogenic culinary disaster. More sophisticated preference satisfaction accounts of welfare take well-being to consist not in the satisfaction of actual preferences, but rather the preferences we would have were we fully informed competent agents. The account still bases our good on our preferences – it is just that our preferences for healthy and tasty food require a set of true beliefs if they are to be actually satisfied. However, the account, while better than the crude version of preference satisfaction theory, remains unsatisfactory.

Informing an agent often matters to an agent's welfare not because it reveals that an object so that it fits our current preferences, but rather because it changes our preferences by pointing to features of the object that make them worthy of being preferred. A friend takes me for a walk in mudflats by the sea for which I have no interest. By pointing out features of the salt marshes I may come to value them a great deal and her informing about the place might make a large difference to my well-being. I now walk through the marshes with developed capabilities to see and hear what is there. But here my well-being is not increased by allowing me to better realise some given preferences, but rather by changes in perception and knowledge to form new preferences. It is for this reason that education, both formal and informal, improves our well-being. Well-being is improved not by satisfying preferences, but by forming our preferences through the development of our human capabilities. The preference satisfaction account of well-being is false.

Once the preference satisfaction account of welfare is rejected, the rationality of refusals of compensation becomes more apparent. Consider an objective account of welfare, say one based on needs understood as the necessary conditions for a person to live a flourishing human life (Wiggins, 1998), or to realise basic functional capabilities (Sen, 1987, 1992b; Nussbaum, 2000). On an objective account there is no reason to assume that goods are ubiquitously substitutable in the way that standard welfare economics assumes. Needs or human capabilities to function are plural, and the goods that satisfy one of them are not substitutable for by goods that satisfy others. Consider, for example, Nussbaum's version of the central human functional capabilities, where these are broadly categorised under different headings: life; bodily health; bodily integrity; senses, imagination and thought; emotions; practical reason; affiliation; other species; play; control over one's political and material environment. Consider some of these headings:

> Bodily Health. Being able to have good health, including reproductive health; to be adequately nourished; to have adequate shelter . . .
>
> Practical Reason. Being able to form a conception of the good and to engage in critical reflection about the planning of one's life . . .
>
> Affiliation. Being able to live with and toward others . . . Having the social bases of self-respect and non-humiliation; being able to be treated as a dignified human being . . .
>
> Other species: Being able to live with concern for and in relation to animals, plants and the world of nature.
>
> (Nussbaum, 2000, pp. 78–80)

Each heading defines a space of capabilities to function. Each space can be itself internally plural, including an irreducible variety of capabilities. Each might also be open to being realised in different way – thus there are a variety of different forms that affiliation might take.

A feature of any such approach is that if there is an irreducible plurality to these different human functional capabilities, there is no reason to expect a ubiquitous substitutability between different goods with respect to welfare. It is not the case that for a loss of good under one heading, say bodily health, there is a gain in other, say practical reasons, that leaves the person's well-being unchanged. There is as people say in everyday parlance, no substitute for good health, for good friends, for particular places and environments. A loss in one dimension can be properly addressed only by the provision of goods in that dimension.[13] A person who suffers from malnutrition requires specific objects of nutrition: more entertainment or better housing will be no substitute. And returning to the example of resistance with which we started, individuals facing eviction from the place in which the life of their community has been lived for generations, facing the disintegration of that community as their homes are flooded for a dam, can properly respond by saying that there is no good that can compensate for that loss. The loss of basic goods in the dimension of affiliation cannot be compensated for by a gain in other dimensions. This is not to say that it would not be better that those who leave should receive something in compensation. However, the idea that there is a sum of money that can be offered that would maintain their level of welfare in the manner assumed by the standard economic theory is a myth founded upon a mistaken theory of welfare. The myth also infects other claims made within environmental economics. In Chapter 6 I will look in more detail at the way that it underlies some of the problems in many approaches to sustainability. The assumption of the ubiquitous substitutability of goods that is the basis of weaker conceptions of sustainability is founded on the assumption of a preference satisfaction theory of well-being. In Chapter 2 we will look further at the role the theory plays in some of the standard justifications of monetary valuation and cost-benefit analysis.[14]

2 Managing without prices

On the monetary valuation of biodiversity

2.1 Managing without prices

Environmental managers manage without prices. In the conflict of values and objectives they face on a day-to-day basis, decisions are normally made without appeal to monetary values and for the most part without appeal to any single common measure. There are resource constraints and different groups argue within those constraints for different priorities among competing objectives. But there is rarely an attempt to use the price mechanism. Thus, in forestry management in the United Kingdom, for example, there exist conflicts between different biodiversity objectives. Increasing the diversity of native tree species in forests is in conflict with the aim of protecting the native species of red squirrel, which fares better than the immigrant grey only in conifers; it conflicts also with the protection of the goshawk, which flourishes in spruce plantation. These competing biodiversity considerations themselves conflict with others: landscape objectives, the use value of forests as a timber resource, the historical and cultural meanings that a woodland might have as a place for a community and so on. Those conflicts of values are at present normally settled without a price ever being assigned to red squirrels, goshawks, or broadleaf woodland. They are resolved through fairly messy looking methods of argument between botanists, ornithologists, zoologists, landscape managers, members of a local community, farmers and so on.

However, it is not hard to find in literature and agency reports on managing biodiversity deep misgivings that this is so. For this is not how neoclassical economic theory says it should be. There exists a large gap between the actual practice of management of biodiversity and the ideal practice that is offered by economists. The ideal assumes that if decisions are to be taken in an ideally rational manner then we need methods of monetary valuation that can price them. Hence as we noted in Chapter 1 the attempt to find methods of arriving at a monetary valuation of biodiversity and other environmental goods: hedonic pricing methods in which a proxy good in the market such as property values is employed to estimate a price for environmental goods; time cost methods which employ the costs incurred by individuals to use an environmental amenity to estimate values; and contingent valuation methods in which monetary values are estimated by asking individuals how much they would be willing to pay for a good or accept in compensation for its loss in a hypothetical market. There is a burgeoning literature on the internal difficulties of each of these, in particular contingent valuation, with increasingly sophisticated attempts to resolve them.[1] These I leave aside. I do so because the whole enterprise of seeking monetary valuations is a mistake. While there may be much that is unsatisfactory about existing procedures for making decisions in the management of biodiversity, these do not include the failure to use monetary values. We can manage without prices. We ought to continue to manage without prices.

Why the belief in the need for monetary valuation of biodiversity? In Chapter 1 I noted and criticised one answer to that kind of question: that we need to do so since the absence of markets is the source of environmental problems. In this chapter I want to look in more detail at one set of arguments that lies behind that view which I have already touched upon in that chapter. These arguments proceed from some broadly utilitarian considerations about how we should arrive at rational decisions. The arguments appear in standard advisory reports on environmental management. Consider, for example, the following from a consultation report for the UK Forestry Commission, *The Valuation of Biodiversity in UK Forests*:

> The Forestry Commission is committed to increasing biodiversity in forests, but faces the crucial question of how much? Valuation of biodiversity in theory presents a tool for such trade-offs, permitting the commission to assess *at the margin* the balance between managing forests for timber production, biodiversity and other forestry values. Economic analysis assists in the allocation of scarce resources, allowing the trade off between say 5 units of A for 6 units of B, within the overall objective of maximising society's welfare.
>
> (Environmental Resources Ltd, 1993, 2.1, p. 8, emphasis in the original)

The report refers here to two of the central arguments for monetary valuation offered by its proponents: that it provides a way of arriving at a decision that maximises 'society's well-being' and that it provides a way of trading off objectives. While the two points are run together in the passage quoted, they are logically independent: one can trade off objectives without being committed to the goal of maximising well-being. I will argue in this chapter that neither offers a reason for monetary valuations.

2.2 Maximising well-being

Modern normative welfare economics has its origins in utilitarianism, and retains utilitarian presuppositions. It assumes that economic decisions, policy and institutions are to be appraised by how far they affect the well-being of affected agents. While its founders assumed a classical hedonistic account of well-being, modern welfare economics tends to assume its preference-based relative. Well-being consists in the satisfaction of preferences, the stronger the preference the greater the improvement in well-being. The strength of a person's preference for an object is measured by his or her willingness to pay for its satisfaction at the margin. We can ascertain the total well-being produced by a policy option by measuring the strength of preference of affected parties for or against its realisation by the willingness to pay measures and then aggregating the result in standard cost-benefit analysis. The central problem given for the environmental economist on this view is as we noted in Chapter 1 – to ensure that preferences for environmental goods that are not registered in actual markets are measured and introduced into the utilitarian sums.

The utilitarian background to welfare economics is normally implicit in the literature on environmental economics. Pearce and Moran (1995) make it explicit in taking a robust line in answering critics who claim that environmental economics fails to capture moral or evaluative standpoints that inform environmental concerns, that it is concerned only with consumer preferences, not with environmental values. To this charge Pearce and Moran respond thus:

> The idea that the 'moral' view is opposed to the 'economic' view rests on many confusions . . . [T]he economic view is itself a moral view – it takes what is effectively a *utilitarian* approach to conservation. What critics are complaining of is not so much the economics as the underlying philosophy of normative economics, *utilitarianism*.
>
> (Pearce and Moran, 1995, p. 30, emphases in the original)

Monetary valuation is an attempt to measure welfare gains and losses for the purposes of the application of a particular direct form of utilitarianism to public policy.[2]

However, while this may answer the objection that normative economics does not have an ethical position, that it stays at the level of consumer preference rather than values, it in effect merely shifts the ground of the argument. A preference utilitarian must either treat other competing and (from its perspective) 'false' ethical viewpoints as themselves mere preferences to be aggregated with other 'non-moral' preferences, or more strongly launder out preferences informed by 'false' ethical theory. Environmental economics effectively does the first – non-utilitarian ethical commitments for environmental goods are effectively treated as further preferences to be included in the sums, being priced under the heading of non-use existence value (Aldred, 1996). At this point the problems reappear, for the proponents of competing alternatives can properly complain that the decision to employ a utilitarian perspective itself demands debate.

Ethical 'preferences' are not like non-ethical preferences. They are substantive positions that reject the presuppositions of utilitarianism and the resolution of the differences between them is a matter of public argument, not aggregation. Protest bids to contingent valuation surveys on this account become quite proper response. If 'preferences' for environmental goods are matters of ethical principle then such commitments are exhibited precisely by a refusal to betray them when offered cash (O'Neill, 1993a, pp. 118–122; 1997a). One should no more accept a price where issues of environmental value are involved than one should on issues of abortion, euthanasia, commercial surrogacy, hanging or any other issue of principle (Holland, 1995). To engage in monetary valuation in these arenas would be quite properly rejected as inappropriate, to ask willingness to pay questions an exercise in corruption. The proper mode of resolution is public debate in which utilitarians have to state their case with others – as they now do in issues of abortion, euthanasia, commercial surrogacy or hanging.

To this line of argument there is a standard line of response which makes a more modest claim for the job of economic valuation. The economist presents the utilitarian welfare-maximising option through the measurement and aggregation of preferences for various goods. It is then for the process of political deliberation to consider that option alongside others – to add such constraints of justice, rights and other ethical values to the pursuit of welfare maximisation as is seen proper. Economic valuation becomes just one part of the policy-making process. That position it is often noted corresponds to the actual role economic valuation plays in decision making. No sane economist ever claimed that it could replace the process of political deliberation.

This fall-back position represents a much stronger defence of environmental valuation than does the undiluted appeal to the utilitarian position. However, in the end it also fails. It does so because it is far from clear that the economic valuation is even an appropriate means to finding the welfare maximising option. An initial problem concerns the definition of welfare in terms of preference satisfaction that the position presupposes. For reasons already discussed in Chapter 1, it is implausible to assume that the satisfaction of preferences as such is either constitutive of welfare or leads to an increase of welfare. I might prefer to smoke in the absence of knowledge of its health effects. Had I been fully informed my preferences would have been otherwise, and the satisfaction of raw preferences might lead to a decrease not an increase in well-being. Correspondingly, individuals' preferences change with the information they have: education, walking or climbing in mountain regions, engaging in skilled activities such as carpentry, all serve to transform our preferences. The standard move by the preference utilitarian is to shift the account of well-being from the satisfaction of raw preferences to the satisfaction of fully informed preferences (Griffin, 1986). Well-being consists in the satisfaction of fully informed preferences: the stronger the preference the greater the well-being.

The shift has important consequences for the valuation of environmental goods. It is well recognised in the literature on economic valuation that changes in the quantity and quality of information one presents and indeed the form in which one presents it will alter responses in willingness to pay surveys about environmental goods. Generally the better the information and its presentation the higher the bid. This is of particular significance for the valuation of biodiversity where most of us have little information and even the best informed acknowledge they are, from the point of view of an 'ideal' 'fully informed observer', ill informed. A variety of methods for eliciting informed responses to contingent valuation surveys have been suggested, from attempting to inform the respondents to the use of expert groups.

These responses are unsatisfactory for two related reasons, one practical, the other theoretical. The practical problem is that any valuation becomes an artefact of the survey: the price you get out depends on the information and presentation you put in. This is a consequence of the second theoretical point, which was alluded to in Chapter 1, that both price and preference are irrelevant once one moves to informed preferences. The point is summarised well by Griffin (1986) in his presentation of the informed preference account of well-being:

> What makes us desire the things we desire, when informed, is something about them – their features or properties. But why bother then with informed desire, when we can go directly to what it is about objects that shape informed desires in the first place? If what really matters are certain sorts of reasons for action, to be found outside desires in the qualities of their objects, why not explain well-being directly in terms of them.
>
> (Griffin, 1986, p. 17)

Thus in the case of biodiversity, to introduce information is not to provide better grounded beliefs to realise a given set of preferences: it is to alter preferences by pointing out features of an object that make them valuable. One is not increasing a person's well-being by allowing them to better realise their given preferences, but rather by allowing them to realise better preferences. That is what education, both formal and informal, is all about. The reason why the preferences of the informed respondent count is that they are in a better position to make judgements about the value of different habitats. What is important is not any price that is put upon the habitats but the soundness of the information and reasons a person has for valuing the habitat. What matters in the valuations is not the preference, but the quality of reasons and information. What is presented as an exercise in eliciting monetary valuation is being transformed into an occasion for educating the respondent. Moreover, this is how it should be. Improvements in well-being come through public deliberation and education of our preferences, not simply through satisfying those we have (for an elaboration of these arguments see O'Neill, 1993a, ch. 5; 1998a, ch. 3).

2.3 Trading-off values

Defenders of economic valuation often retreat from the robust utilitarian position to a weaker claim about the need to find a procedure to resolve conflicts between different objectives in the management of a habitat. The process of monetary valuation, it is claimed, gives one a common measure through which one can trade off different objectives, and, as I noted above, one can trade off objectives without being committed to the goal of maximising well-being. Thus to the philosophical objections to monetary valuation the report for Forestry Commission responds thus:

> The divide on the issue of monetary valuation is profound; but we would argue that some form of prioritisation and mechanism for trade-off is necessary . . . As soon as the concept of trade-off is recognised, then the relevance of some form of valuation is clear. Monetary valuation may not be perfect, nor even possible; but the case for monetary value rests on the fact that it is easier to

> interpret explicitly the meaning of imputed values if they are expressed in a common medium of exchange.
>
> (Environmental Resources Ltd, 1993, 3.4.4, p. 20)

The claim here is that given the existence of competing objectives – biodiversity, landscape, timber, cultural meanings, historical and scientific values and so on – resolution requires some common measure of comparison for giving each its due against each other and monetary price is the best measure. Environmental economists sometimes add that this is just a matter of establishing a common measure. If one wanted to do it in spotted owl equivalents that would do. Money, as the universal medium of trade, is simply easier to comprehend. Thus goes the response.

One standard line of reply to this defence is to deny that all objectives are tradable. The standard and not altogether helpful language for blocking tradability in recent ethical discourse is that of rights. Individuals have rights, for example not to be tortured, and those rights are not open to be traded against increases in human welfare. Where conflicts exist, rights are trumps (Dworkin, 1977, p. xi). To this the strong response from the environmental economist is that choices have to be made. If this is the case absolute prohibitions expressed in the language of rights are inappropriate. We need some ranking of goods.

> If all biological resources have 'rights' to existence than presumably it is not possible to choose between the extinction of one set of them rather than another. All losses become morally wrong. But biodiversity loss proceeds apace . . . [I]t is essential to choose between different areas of policy intervention – not everything can be saved . . . If not everything can be saved then a *ranking* procedure is required. And such a ranking is not consistent with arguing that everything has a right to exist.
>
> (Pearce and Moran, 1995, p. 32, emphasis in the original)

There is much in this response which is right. We face real and unavoidable practical dilemmas in which different values pull us in different directions. And choices do have to be made. The attempt to introduce rights as trump values does not resolve the issue either in the environmental arena or in the arena of human conflicts. The trump values themselves are open to conflict. Given irreducible value plurality, practical dilemmas are unavoidable.

The problem with the position of Pearce and Moran (1995) and other defenders of monetary valuation lies not in the claim that we have to make choices in contexts in which different values pull in different directions, but rather in the picture of what rational choice has to be like in such contexts. It is assumed that rational choice in such contexts requires us to come up with some single common measure which we can use to score each of the different alternatives such that we can rank them by means of some algorithmic procedure. Money as the universal equivalent is taken to provide the most appropriate measure to render commensurate different values. That picture of rational choice is mistaken. Different values are incommensurable: there is no unit through which the different values to which appeal is made in managing a particular site can be placed upon a common scale. Given conflicts between landscape values, biodiversity, timber use, cultural values, there is no substitute for good practical judgement that is informed by debate among practitioners and citizens.

The use of money prices does raise particular problems of its own. As I noted in Chapter 1, there is a special incommensurability between ethical commitments and monetary values. One demonstrates ethical commitments by the very refusal to accept a price upon them: to accept a price is an act with a social meaning – an act of betrayal. These are special problems with monetary units. Moreover, there are also special problems in treating environmental goods as bounded and demarcated commodities over which property rights can be defined (Vatn and Bromley, 1994, p. 137). With environmental goods, in particular complex goods like biodiversity, it is not possible to demarcate them in that way and treat them as commodities.

However, while the use of money values raises particular problems, the problem lies not just in the use of the money measure, but in the picture of practical reason assumed. The more general assumption that underlies the pursuit of a common measure – that we need a measure to make choices into exercises in mathematics – is a mistake. This is true even of the attempt to produce biodiversity indices prior to any attempt at an economic valuation. Whatever heuristic value scoring systems might have in making explicit and transparent the different factors that go to make up a judgement of the worth of a particular site, they cannot themselves serve to replace judgements. Where they do they disguise the fact that a professional judgement has been made.

The problems that arise are illustrated nicely where the attempt at measurement comes into the explicitly deliberative procedures of the legal inquiry. Consider the study by Yearley (1989) of the fate of the ecologists' scoring system for ranking the worth of Irish peat bogs under the scrutiny of a barrister in a public inquiry on horticultural peat extraction. The barrister for the developer effectively criticised scientists for both disguising the judgements they had made about the weights of the attributes of peat bogs – naturalness, the variety of plant species, and so on – and the selectiveness of what was included, for example, the fact that it supported a rare butterfly was mentioned in the inquiry but failed to appear in the scoring system (Yearley, 1989). The criticisms had power because they apparently undermined the 'objectivity' of the scores. The problems here do not lie with the particular features of the scoring system, nor with the ideal of objectivity. They lie rather in the pretence that what is in fact a process of judgement can be replaced by an algorithmic procedure.

There is no unit by which the values of naturalness and variety can be compared, or the value of the red squirrel and the variety of native British trees can be placed on a common scale. Choices here are ultimately a matter of argument, deliberation and judgement. That this is the case is not a matter for regret as a departure from rationality and objectivity. It is what is constitutive of rational choice. The ideal of choice through sums not only is not possible, but also is not an ideal. That actual practice falls short of it is not a matter of regret about practice, but pause for reconsideration of the neoclassical economists' ideal. At both the micro level of managing particular sites and at the macro level of public policy, the view that one could or should replace reasoned dialogue and the judgements of citizens and practitioners by algorithms over numbers is a mistake. While there is plenty of scope for improving the conditions in which reasoned dialogue can occur, there is no scope for the suggestion we replace the public use of reason with mathematical recipes.

2.4 Pragmatic justifications

There is a final line of defence for the practice of economic valuation which shifts the debate from the theoretical ground I have outlined thus far to 'the standpoint of getting things done' (Pearce and Moran, 1995, p. 30). The defence goes that if you want to be effective in protecting the environment then you need to come up with some prices for environmental goods. There are two lines of arguments that are given for this claim: first, that the cause of the loss of biodiversity is the absence of prices for their benefits, and second, that the policy-making community will only ever listen to arguments stated in terms of money values.

The first argument has its source in neoclassical economic theory. Environmental problems like those involved in the loss of biodiversity are the consequence of 'market failures' that arise when real markets depart from 'ideal markets' which, according to the fundamental theorem of neoclassical economics, yield Pareto-optimal outcomes. Of particular significance is that the external benefits of biodiversity and other environmental goods are not priced in the market (Pearce et al., 1989, pp. 5ff.; Pearce and Turner, 1990, p. 41; Pearce and Moran, 1995, chs. 3–4). Hence, the most effective way to respond to the problems is to 'place proper values on the services provided by natural environments' (Pearce et al., 1989, p. 5), services which at present come free of charge. So goes the argument. The argument assumes a norm of market determination of the production and consumption of goods founded on Pareto optimality criteria. There are those who think that the status of this norm itself is in need of critical scrutiny – that Pareto-optimal outcomes are not optimal social outcomes. However, even given these neoclassical assumptions the response of Pearce and others fails. It plays on an ambiguity in the notion of putting a proper price on a good. There are actual prices that are paid in the marketplace, changes in which alter the behaviour of agents. And there are theoretical prices constructed by economists. Theoretical prices constructed by economists have no power in actual markets. They are rhetorical devices in arguments with governments to persuade them to intervene in markets to protect environmental goods.

Consider Pearce and Moran's (1995) economic survey of 'non-use' or 'existence' values of individuals for wildlife (Table 2.1). They offer the following comment on Table 2.1:

> While we cannot say that similar kinds of expressed values will arise for protection of biodiversity in other countries, even a benchmark figure of, say,

> $10 pa per person for the rich countries of Europe and North America would produce a fund of $4 billion pa. This is around four times the mooted size of the fund that will be available to the Global Environmental facility in its operational phase as the financial mechanism under the two Rio Conventions.
> (Pearce and Moran, 1995, pp. 39–40)

The argument puts no real price on environmental benefits. The benchmark price arrived at serves as a rhetorical tool, and by any standards a very blunt and primitive one, to get governments to produce more money. But this is the real point. The claim is that from 'the standpoint of getting things done', it is the only tool around to change policies.

The first pragmatic argument turns out to be in effect a version of the second which constitutes the final fall-back position in defences of environmental valuation. If you want influence, talk money. That is the language that bureaucrats, voters and politicians listen to and understand. It is not enough to establish through argument and debate that the preservation of biodiversity is important. To have an effect one needs to express its importance in monetary terms. This argument is influential among not only believers in economic valuations, but also non-believers who don't really believe it makes any real sense. The problem is to save biodiversity and if that requires the creation of some imaginary prices for the benefits of

Table 2.1 Preference valuations for endangered species and prized habitats

	Species	*Preference valuations (US 1990 $ pa per person)*
Norway	brown bear, wolf and wolverine	15.0
USA	bald eagle	12.4
	emerald shiner	4.5
	grizzly bear	18.5
	bighorn sheep	8.6
	whooping crane	1.2
	blue whale	9.3
	bottlenose dolphin	7.0
	California sea otter	8.1
	Northern elephant seal	8.1
	humpback whales	40–48 (without information)
		49–64 (with information)
Habitat		
USA	Grand Canyon (visibility)	27.0
	Colorado wilderness	9.3–21.2
Australia	Nadgee Nature Reserve NSW	28.1
	Kakadu Conservation Zone, NT	40.0 (minor damage)
		93.0 (major damage)
UK	nature reserves	40.0 ('experts' only)
Norway	conservation of rivers	59.0–107.0

Source: Pearce and Moran, 1995, p. 40

biodiversity then that is what we should do. Thus when Pearce and Moran (1995) come up with fabulous numbers for biodiversity this might not be theoretically defensible, but that is what is required in the policy-making game. It is part of the policy-making ritual to come up with financial figures even if one does not believe them since they are unbelievable.

This is the pure pragmatist position. Where it is used by the non-believer like all similar forms of pragmatism it is open to the charge that it shows a lack of integrity on the part of the environmentalist. It is to collude in a policy-making community which is held together by a common fiction that, a few true believers aside, all know to be a fiction. It is to live a lie. Against that one might invoke an ethic of integrity that insists to use a phrase of Havel's that one lives in truth (Havel, 1989). The standard pragmatist response to that move is a consequentialist one. There are things that are more important than personal integrity not least preventing the destruction of the environment. One needs to subordinate integrity to the greater good. To fail to do so is simply moral squeamishness, a refusal to get one's hands dirty in the pursuit of the common good. Now I believe that this consequentialist reasoning should be resisted (O'Neill, 1995a). However, there are grounds for rejecting monetary environmental valuation from consequentialist premises.

It is far from clear that in the long term, coming up with unbelievable financial figures is the best way of protecting biodiversity. This is not just because the figures are unbelievable. The very treatment of the biodiversity in terms of commercial norms itself is part of our environmental crisis. It is because, as I argued in Chapter 1, the issue of environmental policy is, in part, one of market boundaries. There is a general tendency in the modern world for the domains of commerce to expand: a series of non-market goods such as the human body, especially the womb, academic knowledge, libraries, educational and cultural goods, political deliberation and personal relationships are being subject either to direct commodification or the introduction of market norms. The boundaries that separate the 'free' unpriced world of knowledge, the body and so on from those of the market are being eroded. The appropriate response to the erosion of such boundaries is not to make sure that, as they disappear, the best price is achieved. It is rather to resist the disappearance of the proper boundaries between the different spheres. For example, it is neither a morally nor pragmatically adequate response to commercial surrogacy to work out good commercial rents for wombs, rather than resist commercialisation. The same is true of environmental goods. It may be the case that the environment is unpriced and in a world in which market norms predominate this might be a problem. But strategically it is a move in the wrong direction to accept the disappearance of boundaries and simply look for a price. Protection of our environment is best served, not by bringing the environment into a surrogate version of the commercial world, but by its protection as a sphere outside the world of commodity exchange and its norms. We best serve environmental goals by resisting the spread of market norms.

To thus defend the practice of managing without prices is not then to say that the current ways in which the environment is managed is beyond question. To criticise methods of monetary valuation is not to endorse existing deliberative institutions. The deliberative institutions that are currently used in making environmental decisions clearly have problems. They exclude from the process of deliberation many who have not only an interest in the decision but forms of knowledge and capacities for valuation that are important for arriving at defensible decisions. The public inquiry as a deliberative institution itself is often a ritual through which 'development' is legitimised (Wynne, 1982). However, these and other problems are not primarily about the technical instruments of management. They are about power, the economic structures that sustain certain paths of economic development, the disappearance of public spaces for deliberation and other larger social dimensions to the current environmental crisis. The debate needs to move on from the criticism of economic methods of valuations to consideration of the nature of proper deliberative institutions for resolving environmental problems and of the social and economics framework that will sustain these. In the final part of this book I turn to that project.[3]

3 Property, care and environment

3.1 Property rights and the environment

I opened Chapter 1 with the question 'What is the source of our environmental problems?' And I noted that one influential answer to that question from within neoclassical economic theory is that they have their source in the absence of markets for environmental goods. Preferences for environmental goods are not registered in market transactions. Environmental problems are typically understood as 'market failures' due to externalities or the existence of public goods. If the explanation of environmental damage is that preferences for environmental goods are not revealed in market exchanges, then the solution is to ensure that they are. To bring environmental goods into the market exchange requires the definition of property rights over them such that they can in principle be exchanged. Hence the other formulation of the explanation of environmental problems that is offered within both neoclassical and Austrian traditions – that environmental problems stem largely from a lack of well-defined property rights in environmental goods. Define property rights over the goods, allow them to be exchanged in markets, and preferences for environmental goods will be registered in market exchanges. Where it is not possible to bring environmental goods into actual markets through an extension of tradable property rights, we should construct shadow prices for environmental goods by ascertaining what individuals would pay for them were there a market.

Economic valuation either hypothetical or actual requires that environmental goods be understood as commodities which could in principle be exchanged in markets. As such valuation presupposes that environmental goods can be brought under not just any set of property rights, but a particular set of rights: a liberal conception of property or full property rights are required for market exchange. The development of market economies requires property rights that involve exclusivity and alienability. As Macpherson (1973) notes:

> property as exclusive, alienable, 'absolute', individual, or corporate rights in things was required by the full market society because and in so far as the market was expected to do the whole work of allocation of natural resources and capital and labour among possible uses.
>
> (Macpherson, 1973, p. 133)

Full liberal rights require that the bundle of rights, duties and liabilities that constitute ownership – rights to possess, use, manage, income, transmissibility, liabilities for compensation for injuries and so on (Hohfeld, 1913) – are formally assigned to a particular agent, be that an individual or corporate. The tendency to concentrate such rights and liabilities is a condition of their alienability in the contractual arrangements of market exchange. In contrast, rights and liabilities in pre-modern property systems are often diffused rather than concentrated, and customary rather than contractual.[1]

On one model then environmental problems require for their solution the definition of full liberal property rights over goods that will enable them to register their value in actual or hypothetical markets. How adequate is that solution? I have already given some reasons to question that view in chapters 1 and 2. In this chapter I give further reasons for scepticism. I will do so by placing recent liberal arguments in the context of older debates about property, in particular those concerned with the distribution of care. While proposals for the extension of liberal property rights over environmental goods often appeal to arguments from the need to distribute care, I show that there are conflicts between them. Care for particular places that embody the life of a community that has an existence over time is often expressed through resistance to liberal property rights. We express mutual obligations to members of a community through a denial of exclusive property rights to particular individuals or corporations over certain common goods. Finally, what constitutes care for environmental goods itself is contested across class, occupation, culture and history. Conflicts between those with different conceptions of care are often expressed through conflicts in property rights. From Locke to the present, appeal to specific conceptions of care to justify property has been invoked by powerful interests to legitimise the appropriation of goods of those whose activities embody a different conception of care. The introduction and maintenance of liberal property rights regimes involves the creation and sustenance of a particular distribution of social power and should be understood as such.

3.2 Care, property and community

Some of the apparent plausibility of the very idea that the definition of property rights might be a proper response to environmental problems stems from a more general argument for private property that is much older than market economies. It is an argument from the distribution of care and responsibility. The argument goes back at least as far as to Aristotle's response to Plato's defence of communism and the abolition of the family in the *Republic*. Aristotle's response to both communism and the abolition of the family has the same theme – that a special relation to particulars is a condition for care for them:

> What is common to the greatest number gets the least amount of care. Men pay most attention to what is their own: they care less for what is common; or at any rate, they care for it only to the extent to which each is individually

> concerned. Even where there is no other cause for inattention, men are more prone to neglect their duty when they think that another is attending to it.
>
> (Aristotle, 1948, II.iii)

Just as care for particular other individuals, expressed through kinship relations, is preferable to generalised concern – 'better a real cousin than a Platonic brother' – so also care for particular objects and places is preferable to the generalised concern involved in Plato's version of communism.

Aristotle's argument has been influential. It runs through medieval debates on the problem of the defensibility of personal property given the biblical assumption that the world was given to the human race in common. Consider for example Aquinas' response to the problem of how private property is justifiable if 'according to the natural law all things are common property'. Aquinas' response echoes Aristotle's claims in the *Politics*:

> [E]very man is more careful to procure what is for himself alone than that which is common to many or to all . . . [H]uman affairs are conducted in more orderly fashion if each man is charged with taking care of some particular thing himself, whereas there would be confusion if everyone had to look after any one thing indeterminately.
>
> (Aquinas, 1975, II.II, 66 a.2)

As we shall see, the Aristotelian argument remained influential on early modern discussions of commercial society. And, more problematically, it is often claimed to be a precursor of recent arguments for the extension of liberal property rights to environmental goods. For example, Aristotle's argument is sometimes cited, mistakenly, as a predecessor to Hardin's influential 'tragedy of the commons' (Baden and Noonan, 1998, p. xv).

There is, I think, a moment of truth in Aristotle's argument which has important environmental implications. With some qualifications to which I return below, it is true that the maintenance of the quality of land and water, the nurture of resources, the upkeep of landscapes often requires that there be particular persons who have a special interest in particular places and goods. The distribution of care matters. I do not think it follows that property should be private – nor indeed does Aristotle say that. In the first place care needn't be expressed through a property relation at all. I don't own my children, but I do have a special relation to them, and with that particular concerns of care. Similarly particular places and landscapes matter to individuals in the absence of any direct property relations. Consider the following quote from an interview with a person talking about the value of the Yorkshire Dales:[2]

> This is my community if you like. It may sound a strange thing to say, this is my valley. But this is where I am, where I come from. It's what makes me tick if you will.

The possessive pronoun of the quote is a pronoun of identity and not of ownership. You need not own anything in a place to say things like this. Moreover what is being expressed here is a relationship to the landscape that is communal rather than individual. The landscape, the valley, embodies a communal life. Care need not be expressed through private property.

This is not to deny that identity can be expressed through property, a point that lies at the centre of Hegel's defence of private property (Hegel, 1952, paras 34–71). However, against the Hegelian focus on individual personality expressed through private property, common property and public goods are also ways of expressing a communal identity. Moreover, one source of resistance to the spread of liberal property regimes and attempts to price environmental goods is the way that it undermines the expression of relations of community and of a communal identity, in common goods. Consider again the protest I discussed in Chapter 1 to a request to price the environment in the Narmada Valley to be drowned by a dam. Here again is part of that passage:

> How are you compensating us for fields either – we didn't buy this land; our forefathers cleared it and settled here. What price this land? Our gods, the support of those who are our kin – what price do you have for these? Our adivasi (tribal) life – what price do you put on it?
>
> (Bava Mahalia, 1994)

One central point in this passage is that the life of a community, both to contemporary kin and over time to those who have gone before and settled the land, cannot be expressed in monetary prices. The land embodies relations to a community, a way of life, that is not of a kind that can be captured in alienable liberal property rights. Moreover, it is as an embodiment of the life of a community that care is shown for it. Thus while it may be true that some impersonal common property regimes can be such that no one cares for particular goods – bureaucratically managed property regimes in particular are prone to that failing – there are well-functioning common property regimes that themselves successfully distribute care and responsibilities over different goods (Ostrom, 1990; Bromley, 1991, ch. 2). More generally, the diffused nature of rights and responsibilities in pre-modern economies is compatible with Aristotle's position. Aristotle, in defending private property against Plato, was not defending full liberal ownership – the institution did not exist and the system of ownership he defends is very different.[3]

These points about the relations between community identity and public or common property do not only have relevance to the defence of traditional common property regimes. They have more general implications for arguments about the environment and property in modern societies. Arguments about the boundaries of what can become an object of commercial exchange in modern societies are in part arguments about the nature of social relations. This is evident, for example, in the debates around Titmuss's (1970) defence of non-commercial blood donation, that to make blood an object that can be only a gift, not exchangeable property, is to protect relations of social commitment to others.[4] They are also evident in a form still more clearly relevant to environmental concerns in arguments about the nature of public goods.

3.3 Normative public goods

There is a little noticed ambivalence that runs through many discussions of public goods. Public goods in the economic sense are seen from within the perspective of neoclassical economics to be a problem – as a source of 'market failure'. Public goods are those that are (i) 'non-rival' in consumption – the consumption of the good by one person does not decrease that of others and (ii) 'non-excludable' – individuals cannot be excluded from the benefit of using the good. The claim goes that such goods cannot be adequately provided by the market, since each individual has a reason to be a free-rider, to benefit from the consumption of the good without contributing towards its costs. While there may be no pure public goods, many goods, including environmental goods, approximate to public goods. Two solutions are normally invoked in response to the 'problem' of public goods: state regulation and provision, or the establishment of property rights which do allow excludability.

Public goods in the economic sense are goods from which individuals cannot be excluded. However, there is an ethical and political sense of the term public good that needs to be distinguished here, that is, a good is a public good if it is the case that individuals ought not to be excluded from its use.[5] Call this a normative public good. The economic and normative senses are logically distinct. A good from which individuals can be excluded is not necessarily one from which they ought to be excluded. However, while the senses are distinct, arguments that a certain good is a public good in the economic sense sometimes seem to be used as a surrogate way of defending the public nature of the good which for quite distinct reasons is believed to be a normative public good which ought to be nonexcludable. Consider for example the case of scientific knowledge. It is sometimes claimed that such knowledge is not a good that can be treated as a marketable good because it is a public good, that is non-rival and non-excludable (David, 1992, p. 11). For reasons I develop elsewhere I think this argument is flawed (O'Neill, 1998a, ch. 11). Moreover in practice the establishment of private property rights over the results of research is fast becoming the norm. The case against the spread of this property regime is not that science is a public good that cannot become the object of private appropriation – it too clearly can. It is rather a case that it *ought* not to become such an object for a number of reasons. Property rights in scientific knowledge are incompatible with the open communication in science which is a necessary condition for the growth of knowledge as such. Open communication and the absence of ties to special interests is also a condition for the public credibility of science which is necessary for its role in democratic debate. Finally there are distributive arguments for free access to knowledge: knowledge is a condition of the development of human capacities and as such through free education ought to be open to all. I have defended the normative non-excludability of scientific knowledge in more detail elsewhere (O'Neill, 1998a, ch. 11). Even if it is not an economic public good, scientific knowledge is a normative public good that defines the relationships constitutive of the traditional scientific community and its relationships to wider societies.

The status of normative public goods applies to a number of goods that are central to recent environmental debates. Indeed, the status of knowledge itself as a normative public good itself has become increasingly important to environmental conflicts. The development of new systems of intellectual property rights stands at the centre of many recent environmental disputes. Consider the arguments about ownership of genetic resources. As with other resources, the standard view in the economics literature and in policy informed by that literature is that the decrease in biodiversity is due to the absence of ownership in genetic resources. By bringing the genetic stock into a clearly specified system of liberal property rights, for example through a system of patenting, the world's genetic stock will be preserved. Against that view stands that which insists that genetic resources in the form of traditional seed varieties have been sustained without such exclusive rights and that these resources are not the kind of good to which exclusive rights are appropriate. They are goods which ought to be common, in the sense of being available to all, since for many they are a condition of human livelihood. Hence, the struggle for seeds takes the form of the assertion of traditional fact of usage and the political and ethical claims made on behalf of such usage against the attempt to assert property rights and with it control by commercial corporations.

Many other environmental goods are at the heart of conflicts about what ought to be common in the sense of available to all in a world characterised by the extension of markets and property of an exclusive liberal kind. Such disputes are in part arguments about what kinds of communities we take ourselves to belong. The goods that any community defines as normative public goods from which members should not be excluded define the relationships of need and mutual obligation that are constitutive of that community. They define what it is to be a member of that community. The arguments about the public provision of health based on need are of this kind. To change health from a public to a private good is to redefine the community of which we are members. Similarly for many environmental goods, from basic livelihood goods such as clean water and unpolluted air to goods necessary for quality of life such as access to open spaces, to define them as goods from which no member ought to be excluded is to define what is owed to members of that community.

3.4 Property, community and care over time

The significance of community and identity for the protection of environmental goods and its problematic relation to liberal property rights is also evident in problems concerning the distribution of care over time. While the appeal to the Aristotelian tradition of arguments from the distribution of care are sometimes called upon in recent defences of liberal property rights, if one examines the debates around land and commerce in the seventeenth and eighteenth centuries one finds the Aristotelian argument employed in the opposite direction, against the development of commercial property in land. Full liberal ownership, the development of exclusive and alienable rights in land, was viewed to be incompatible with intergenerational care. Thus the civic humanist tradition called upon the Aristotelian defence of property ownership against the extension of tradable property rights in land. The civic humanist criticism of commercial society was founded on the belief that the civic virtues had their basis in stable ownership of landed property (Pocock, 1975, chs. 13–14; 1985). The material foundation of a good society lay in 'real property recognizable as stable enough to link successive generations in social relationships belonging to, or founded in, the order of nature' (Pocock, 1975, p. 458). Commercial society, by treating land as an alienable commodity like any other, mobilised the land and hence undermined that link between generations. The material foundation for the proper distribution of intergenerational care was undermined. The argument is echoed in the intergenerational references in the passage about the Narmada Valley quoted earlier. It is also evident in some of the protests to contingent valuation which object to the way that the practice treats environmental goods as if they were alienable property discussed in Chapter 1.

The issue of the relationship between liberal property rights and intergenerational care is an important, if neglected one. Whatever the defensibility of the particular forms of land-ownership the civic humanists defended there is some force in their arguments about the commercialisation of land. The force is apparent if one considers one of the most influential arguments for the extension of liberal property rights to environmental goods – Hardin's well-known and, as he himself now admits, mis-characterised 'tragedy of the commons' (Hardin, 1998).[6] In its original form the tragedy runs roughly as follows: a pasture is the common property of a number of herdsmen, each one of whom aims to maximise his own utility. For any individual, the positive benefit of adding another animal to his herd will be greater than the loss from overgrazing, since the benefit accrues entirely to the individual, while the loss is shared among all the herders. Hence, a set of rational, self-maximising herdsmen will increase their herds even though collectively it is to the detriment of all. Hardin's tragedy of the commons is founded upon the existence of a pay-off structure for each individual that places individuals in a many-person prisoner's dilemma. For each individual the matrix shown in the diagram holds.

	Others graze additional animal	Others don't graze additional animal
I graze additional animal	3rd	1st
I do not graze additional animal	4th	2nd

Given that pay-off structure, each rational agent will increase herd size to the collective ruin of all. Whatever other agents do, the best strategy is to increase one's own herd. Two standard solutions are offered to the tragedy. The first is action through coercion. The second is the privatisation of the commons. For a period 'the tragedy of the commons' was taken to provide an argument for the privatisation of the commons. It is a now well-developed point that Hardin's tragedy is not a tragedy of common ownership at all (Ostrom, 1990; Aguilera-Klink, 1994). Hardin's tragedy is a problem not of common ownership, but of open access in a context of private ownership of particular assets. That line of reply is well known and I leave it aside here. I want to focus rather on the problem of intergenerational care.

A further difficulty with Hardin's argument becomes apparent if one considers the use of resources over time. With respect to the earth, successive generations occupy a temporal 'commons'. This is true even given complete private ownership of resources. To reuse Hardin's example, consider a plot of land owned by successive generations of herders. Each generation, if they are rational self-maximisers, will add to their herd and graze the land to its limits within that generation; the benefit accrues to themselves, while the loss is shared by all successive generations. Hence, given Hardin's logic, we should expect each generation to deplete the resources it passes on to following generations.

What is the solution to this temporal tragedy? Neither of the standard solutions – privatisation or coercion – is available. Thus the standard solution that is offered to the original spatial tragedy of the commons, to place the 'commons' under private ownership, has no application. The result occurs even given private ownership within each generation. Neither is an intergenerational police officer available. No future generation can police the current generation. So what is the solution? Or putting the question differently, why is it the case that successive generations did not always knowingly deplete resources until recently? Indeed, why did they after their own lights sometimes improve them, even given that they knew they would not reap the benefits? Part of the answer lies, as the civic humanists noted, in the conditions that allow individuals to identify with an inter-temporal community over time. Inter-temporal community is often expressed through the denial of full liberal property rights. The expression of obligations to future generations by the denial of full liberal property rights is one that is expressed pithily by Marx:

> a whole society, a nation, or even all simultaneously existing societies taken together, are not the owners of the globe. They are only its possessors, its usufructuaries, and, like *bone patres familias*, they must hand it down to succeeding generations in an improved condition.
>
> (Marx, 1972, p. 776)

The appeal Marx makes here is not to that of common ownership, for all that Marx believed that common ownership in the means of production provides the most justifiable form of ownership within any generation. It is rather to that of usufruct. The appeal to the notion that we are but usufructuaries here is both apt and powerful. To have a usufruct is to have use rights and duties of preservation in an object that is not one's property. Thus the definition in *Justinian's Institutes*: 'Usufruct is the right to the use and fruits of another person's property, with the duty to preserve its substance' (Birks and McLeod, 1987, 2.4.1). To claim that any generation is but a usufructuary is to claim that they have only rights to use and duties to preserve. Rights to alienate through market exchange and exclude later generations from their benefit are denied. Environments are often expressive of social relations between generations. Places embody relations to the past and future of communities to which we belong (O'Neill, 1993a, ch. 3; 1993b). And as we saw in Chapter 1, it is that which in part activates the protests to the demand to express concern for nature in monetary terms.

3.5 In defence of neglect or the meanings of care

In the arguments to this point I have assumed that there is something right in Aristotle's arguments for special relations to particular places that start from the need to distribute care. I claimed that with some qualifications, environmental goods will be fostered if there is a proper distribution of care. The lack of care, neglect, is taken to be bad for the environment. In this section I present the qualifications. Neglect may not be always bad, or at least there is an argument to be had as to the nature of care for the environment. Hardin's argument again offers a useful starting point. Hardin's argument isn't a novel one. One significant anticipation is to be detected in Hume's *A Treatise of Human Nature*:

> Two neighbours may agree to drain a meadow, which they possess in common; because 'tis easy for them to know each others mind; and each must perceive, that the immediate consequence of his failing in his part, is, the abandoning the whole project. But 'tis very difficult, and indeed impossible, that a thousand persons shou'd agree in any such action; it being difficult for them to concert so complicated a design, and still more difficult for them to execute it; while each seeks a pretext to free himself of the trouble and expence, and wou'd lay the whole burden on others. Political society easily remedies both these inconveniences.
>
> (Hume, 1978, III, Part I, section VII)

Hume's tragedy of the undrained meadow anticipates Hardin's tragedy of the commons. Hume's formulation however points to some problems for any view that assumes that the tragedy of the commons is necessarily an environmental tragedy. It is not at all clear that the tragedy of the undrained meadow is an environmental tragedy. Hume's discussion belongs to a set of perceptions and attitudes to nature that are less prevalent than they were.

Discussion of property rights in the seventeenth and eighteenth centuries appeals to a perception of 'unimproved' land as 'waste', which, as such, could be appropriated by others to ensure its proper use. Typical is Locke's comparisons of the 'wild woods and uncultivated waste of America left to Nature without any improvement, tillage or husbandry' (Locke, 1988, 2.37),[7] with the improved and cultivated lands of Britain and his corresponding account of the original appropriation of land:

> Whatsoever he tilled and reaped, laid up and made use of, before it spoiled, that was his particular Right; the Cattle, and Product was also his. But if either the Grass of his Inclosure rotted on the Ground, or the Fruit of his planting perished without gathering, and laying up, this part of the Earth, notwithstanding his Inclosure, was still to be looked on as Waste, and might be the Possession of any other.
>
> (Locke, 1988, 2.37)

Hume shares this perception of unimproved land as waste. The marsh undrained is unimproved.

The dominant perceptions of land and landscapes shifted during the romantic movement. Much recent environmental thought has echoed Mill's romantic influenced observation about the limits of agricultural expansion.

> Nor is there much satisfaction in contemplating the world with nothing left to the spontaneous activity of nature; with every rood of land brought into cultivation, which is capable of growing food for human beings; every flowery waste or natural pasture ploughed up, all quadrupeds or birds which are not domesticated for man's use exterminated as his rivals for food, every hedgerow or superfluous tree rooted out, and scarcely a place left where a wild shrub or flower could grow without being eradicated as a weed in the name of improved agriculture.
>
> (Mill, 1994, Book IV, ch. 6, section 2)

It is that perception that informs modern conservation. In the United Kingdom at least an undrained meadow of the kind that Hume describes would be likely to become the object of a nature conservation order and the farmer be offered payment not to drain the meadow. From this perspective which informs much of modern environmentalism, Hume's particular failure of concerted action is a source of environmental goods. Like many other environmental goods they are the consequence not of purposive human action but of neglect. It would be a mistake to conclude from Hardin's tragedy that the definition of property rights is a necessary condition for the realisation of environmental goods. Any such relationship is at best a contingent relation and requires detailed empirical support and normative judgement, not a priori argument.

This partial praise of neglect might, however, be rephrased as an argument about what constitutes care. The conflicts between the agricultural and Mill's more romantic perceptions are not merely an historical phenomenon. They remain points of conflict between conservation bodies and farmers. Consider the following comments of farmers in the Yorkshire Dales discussing nature conservation schemes in the Yorkshire Dales (Walsh et al., 1996; Walsh, 1997; O'Neill and Walsh, 2000). Farmers maintain a critical attitude to land that is left to 'go wild':

> Plenty of weeds in the fields – that looks a mess, doesn't it . . . A field full of thistles looks a mess. Well it looks nice when it's in flower – a fortnight later

when it goes to seed it is a mess then. And well maintained walls and hedges are better to look at than a wall that's full of holes . . .

F. A farmer will look at someone else's farm and could tell whether it was well farmed or not. They wouldn't look at the view and think – what a good view – they look and see whether it had been well farmed. Yes, they've a different way of looking at it.
M. What would they look for?
F. Tidiness, walls are up, general upkeep of the land. It would look green.

(Walsh et al., 1996)

A farmer will express care for the land through evidence of good husbandry, not through the number of wild flowers it contains. Indeed from the farmer's perspective, to leave one's fields to be overrun by wild flowers is to reveal an absence of care. What counts as 'care' for the farmer will not be what counts as care for the conservation agency.[8]

The conflict between different conceptions of care is often expressed as a conflict of rights. Consider the following passage from Emerson about landscapes:

> The charming landscape which I saw this morning, is indubitably made up of some twenty or thirty farms. Miller owns this field, Locke that, and Manning the woodland beyond. But none of them owns the landscape. There is a property in the horizon which no man has but whose eye can integrate all the parts, that is, the poet. This is the best part of these men's farms, yet to this warranty-deed give no title.
>
> (Emerson, 1983, p. 9)[9]

Emerson captures a conflict between well-defined legal property claims in environmental goods and ethical claims that are expressed in terms of the denial of property rights. Landscapes are public spaces in the sense of being spaces in which different communities and individuals dwell. Whether or not they are strictly public goods in the economic sense is I think a moot point. While in certain conditions perceptions of a landscape may be non-rival – my enjoyment of the scene may not always affect those of others – there is a clear sense in which access to the 'scene' or to the 'view' can be controlled. Landscapes are however public in the sense of being shared spaces and these spaces may matter to those who dwell in them in different ways which are not always consistent. As such they are always potential sites of conflict.

Those conflicts are often expressed in terms of conflicts between private ownership and public access that Emerson (1983) captures well. On the one hand there is a very clear sense in which landscapes can be and are objects of ownership in the legal sense. As Emerson notes, rural landscapes are often made up of the properties of particular persons – Miller owns this field, Locke that, and Manning the woodland beyond. In discussions of their attitudes to visitors farmers are often keen to assert those rights. Hence the following from a farmer in the Yorkshire Dales:

> I've said to the National Park. I said you were wrong in the first instance calling it National. It should never have been called National because the people that come up there and if I say anything to them if they're on my land, they say it's National Park, it belongs to us, it doesn't belong to me, it belongs to them because it's called National Park . . . We have rights surely.
>
> (Walsh et al., 1996)

Here is another comment from farmers from the Pevensey Levels in southern England on townspeople visiting the countryside: 'they don't realise that we own the land. They think it's national heritage and they can go where they like' (Burgess et al., 1995, p. 15).

However, interviews with those visitors uncover quite opposing perceptions. Many members of the public expressed strong Emersonian views that reject the supposition that landscapes could be owned:

> Ray: The wildlife that lives on the farmer's land doesn't give ownership to anyone. And you have the right to have a look at the wildlife and to see the natural scenery around so that if you've got a house overlooking farmland and someone sells it and puts some edifices up you object. Your view is spoiled. So we've got those sorts of rights I think which should be preserved in the system . . . If people want to do that [pay for facilities] fine but I don't think I want to pay to just go and see a piece of countryside. As far as I'm concerned, that countryside belongs to everybody and should be accessible to everybody.
>
> (Burgess et al., 1995, p. 83)

The assertion here expresses an attitude to landscapes as normative public goods. Access can be controlled. It ought not to be controlled, because such goods are not the kind of things to which exclusive rights are appropriate. They belong to the poet in everyone.

The existence of such conflicts points to deep problems in the attempt to define property rights as the means to realise 'efficiency'. It is a well-rehearsed theoretical point that the common treatment of efficiency as if it were logically independent of distribution is at best misleading, for the determination of efficiency already presupposes a given distribution of rights. For a given initial distribution of rights, one can derive a Pareto-optimal outcome, but Pareto optimality is always relative to an initial starting point, and cannot tell us what that should be. Different property right regimes are themselves not Pareto comparable. If property rights are changed so also is what is efficient. Hence, the opposition between distributional and efficiency criteria is misleading. Existing costs and benefits themselves are the product of a given distribution of property rights. Since costs are not independent of rights they cannot guide the allocation of rights. Different initial distributions entail differences in whose preferences are to count. Environmental conflicts are often about who has rights to environmental goods, and hence who is to bear the costs and who is to bear the benefits. Policy choices often have significant consequences for the distribution of the rights and incomes of affected parties. Where this is the case one cannot treat distributional issues as a distinct item to be treated *ex post* once efficiency has been met. Hence, environmental policy and resource decision-making cannot avoid making normative choices which include questions of resource distribution, and the relationships between conflicting rights claims (Samuels, 1972; Schmid, 1978; O'Connor and Muir 1995).

The point is reflected in the conflicting claims of rights to landscapes just noted. Consider the responses to the contingent valuation of the Pevensey Levels. The existence of conflicting rights claims has effects on willingness to pay responses, for different beliefs as to who has rights can entail different views as to who is supposed to pay for the maintenance of a particular landscape. Given priority to the farmer's property rights claims, on many standard economic models it follows that a decision that affects the landscape requires compensation for the farmer's losses, the non-rights holder the obligation to pay. However, if the rights to the landscape are prior, then the roles are reversed. The obligations lie upon the farmer, the rights with the visitor or non-farming dweller. However, who has prior rights is often just what is at issue in environmental conflicts. Hence as one respondent to the contingent valuation puts it:

> If they were farming the land rotten, then isn't that the same as a big chimney poking out loads of environmentally, polluting air? Isn't it the same as ICI dumping chemicals in the river? If they're doing it, why are we paying them?
> (Burgess et al., 1995, p. 85)

The issue of willingness to pay is caught up in a much larger social issue of competing views of rights to the countryside, and who has rights to expect compensation for producing an environmental good and who has the duty to pay. Any answer to that question is an issue of politics and ethics. It is not one that economic analysis and economic tools such as cost-benefit analysis are even in theory capable of resolving. Their application assumes that an answer has already been given.

Any claim to the Pareto efficiency of markets has always to be read, 'given this distribution of property rights, this is efficient'. Unqualified statements employing the notion of Paretian efficient outcomes are elliptical. This dependency of efficiency claims on prior distribution of rights is normally implicit and tends to disappear in policy recommendations, be these recommendations for direct markets solutions to environmental policies through the extension of tradable property rights or those that employ surrogate market procedures. Hence, in practice, both tend to a conservatism by assuming a status quo distribution of de facto powers and rights that is actually or potentially contestable, but which in the policy recommendation is rendered invisible. The subsequent solutions have strong distributive impacts. The monetary value of a 'negative externality' depends on social institutions and distributional conflicts – willingness to pay measures, actual or hypothetical, consider preferences of the higher income groups more important than those of lower ones. If the people damaged are poor, the monetary measure of the cost of damage will be lower – 'the poor sell cheap' (Martinez-Alier, 1994; Martinez-Alier and O'Connor, 1996).

The distribution of property rights is not just or even primarily in the modern world about the distribution of care. It is about the distribution of social power. The employment of market solutions to environmental problems or of surrogate market-pricing for the purposes of cost-benefit analysis both presuppose specific distributions of property rights and de facto powers to enforce such rights. Where new property rights regimes are introduced they are defined by those with economic and social power. To invoke 'market solutions' is to invoke a particular distribution of power to determine outcomes. It is not an appeal to a neutral mechanism for the optimal satisfaction of preferences, where the only problem is that of ensuring that preferences for environmental goods are included. Market solutions are mechanisms for defining and defending particular distributions of social power and should be understood and contested as such.[10]

4 Public choice, institutional economics, environmental goods

4.1 The challenge of public choice theory

From a neoclassical perspective the source of our environmental problems lies in the fact that environmental goods are unpriced. The solution to our environmental problems on this account is to put a price on goods. In Chapter 1 I noted that there were two approaches to resolving the problem. The first is to attempt to discover what the price of the good would be were there a market. That price could then enter a standard cost-benefit analysis. The second approach is to create property rights on environmental goods thus allowing them to be traded on actual markets. In Chapter 1 my arguments focused on the first of these approaches, as has a great deal of the literature on economic approaches to environmental decision-making. However, the use of cost-benefit analysis has its opponents within mainstream economics who are still committed to market-based solutions to environmental problems. These include not only the Austrian adversaries of the neoclassical paradigm, but also from within the neoclassical paradigm, the proponents of public choice theory.[1] Whatever the influence of cost-benefit analysis as a justificatory ritual within public policy-making processes, some of the more rigorous theoretical work within neoclassical economics represented by public choice theory is sceptical of its theoretical justifications and dismissive of it in its practice. Moreover, many of the criticisms it makes of cost-benefit analysis are equally applicable to arguments of its opponents who appeal to the ideal of public deliberation.

From the public choice perspective, the central weakness of cost-benefit analysis as a tool for public decision making is its assumption of benign state actors. Cost-benefit analysis is a tool employed by bureaucrats, who never, in practice, leave the office and who impute preferences to those affected by a policy. The results of cost-benefit analysis are then employed by politicians to justify their policies. The defender of cost-benefit analysis assumes that both bureaucrats and politicians are benevolent actors concerned to realise the common good or welfare of all (Olson, 1965, p. 98). However, making that assumption entails that the axioms that define the rational actor in neoclassical theory cease to apply behind the office doors of the bureaucrat or politician. The actors are no longer taken to make rational choices that maximise their own utility: they rather become altruistic channels through which the maximisation of the general utility is achieved. The axioms of neoclassical theory are assumed not to apply in the non-market setting of politics:

'the conventional wisdom holds that the market is made up of private citizens trying benefit themselves, but that government is concerned with something called the public interest' (Tullock, 1970, p. v).

The starting point of public choice theory is the denial of that assumption. There is no reason to assume that what is true of actors in the market ceases to be true when they enter non-market situations. Thus Buchanan writes: '"Public Choice" . . . is really the application and extension of economic theory to the realm of political and governmental choices' (Buchanan, 1978, p. 3).[2] The axioms that characterise the rational agent in economic life should be taken to apply also to the explanation of the behaviour of bureaucrat and politician in their political activities. There is no reason to assume that state actors suddenly become different and more benign when they enter the arena of government. If it is true that individuals act as rational self-interested agents in the marketplace, 'the inference should be that they will also act similarly in other and nonmarket behavioral settings' (Buchanan, 1972, p. 22). The assumption made in neoclassical defences of cost-benefit analysis of benign state actors represents a failure of theoretical rigour and nerve to apply consistently their axioms defining the rational actor.

For public choice theories the public use of cost-benefit analysis does not and could not produce the optimal outcomes of 'ideal markets' by other means. The state actors act to maximise their own interests not the 'public interest'. Bureaucrats are taken to aim at maximising the size of their bureau budget, since that is correlated with their utility:

> Among the several variables that may enter the bureaucrat's motives are: salary, perquisites of the office, public reputation, power, patronage, output of the bureau, ease of making changes, and ease of managing the bureau. All except the last two are a positive function of the total budget of the bureau during the bureaucrat's tenure . . . A bureaucrat's utility need not be strongly dependent on every one of the variables which increase with budget, but it must be positively and continuously associated with its size.
>
> (Niskanen, 1973, pp. 22–23)[3]

Likewise, in explanation of the behaviour of voters and politicians it is standardly assumed that voters act like consumers and political candidates like firms. Candidates aim to maximise votes and hence gain political office, voters to maximise the satisfaction of preferences for those goods the state can deliver.[4] Once economic theory is applied to politics, the state no longer appears as a beneficent representative of the public good. Rather, it is argued that the self-interested behaviour of bureaucrat, politician and voter lead, if unchecked by institutional reform, to the constant expansion of government expenditure and provision, producing outcomes that are irrational and inefficient. 'Market failure' gives way to 'government failure' (Niskanen, 1971; Buchanan 1975, ch. 9; Buchanan and Wagner, 1977; Brennan and Buchanan, 1980; Wolf, 1987).

Public choice theory is consistent with and indeed provides one explanation for an important and under-discussed feature of the use of cost-benefit analysis in the evaluation of public projects: that cost-benefit analyses of public project systematically underestimate the actual costs of projects and overestimate their actual benefits. For example, Flyvbjerg et al. (2002) note in a study of the use of cost-benefit analyses in the transport sector, the actual costs of projects was on average 28 per cent higher than projected costs. This overestimation is a global phenomenon, although more pronounced in developing countries. The overestimation is consistent over time; since the mid 1930s there has been no adjustment of projected costs in the light of previous underestimation of previous costs. There are a variety of explanations one might offer for that phenomenon. However, the public choice theorist can argue that the facts are consistent with public choice theory. This is as one might expect things to be, given self-interested bureaucrats and politicians pursuing their own interest in advancement and power.[5]

The public choice theorist typically appeals to a free market economic policy which attempts to rectify market failure, not by using the state to realise efficient outcomes by bureaucratic means, but by institutional changes within the market (Buchanan and Tullock, 1962, ch. 5). 'Government failure' is thereby avoided. The public choice response to environmental problems that arise from 'market failure' – externalities, public goods and the absence of a market price on many environmental goods – has, then, been to find a solution within the market sphere itself. Direct government intervention is not required to solve problems of market failure. Rather, they can be resolved by a redefinition of property rights within the market. Thus, Coase's theorem (Coase, 1960) is invoked to resolve the problem of externalities: given perfect competition and the absence of transaction costs,[6] solutions to negative externalities, for example pollution, are possible through a process of bargaining, if property rights are properly assigned either to the 'damaging' agent or the 'affected' agent. If the damaging agent has the rights, then the affected agent can compensate him not to continue the damaging activity; if the affected agent has property rights, the damaging agent can compensate him or her to bear the damage. Thus, for example, in the former case a pollution sufferer might compensate the polluter, in the latter, the polluter might compensate the sufferer. Similarly, where unpriced public goods such as clean air and water exist, the optimal solution is not to place a shadow price on the goods, but to define property rights, if not directly over them, then over their use, for example, through pollution permits. Thus tradable pollution permits which allow markets in pollution are defended on the grounds that they address the interests of the actors directly, and hence do not make unrealistic demands on conscience or law; that they encourage pollution to diminish where it is cheapest for it do so; and that they even allow those with preferences for non-pollution to express those preferences directly within the market (Dales, 1968; Anderson and Leal, 1991). The problems of environmental damage consequent on 'market failure' can be resolved not by treating environmental goods *as if* they had a price, but by directly bringing within the realm of market contract. The problems have a solution within the sphere of voluntary market exchange and without recourse to government intervention that leads simply to 'government failure' worse than the failures it is supposed to cure.

4.2 What is dead and what is living in critiques of orthodox environmental economics?

How should the critic of orthodox market-based approaches to environmental problems react to this theoretically more robust defence of the pricing of environmental goods? How much of the criticism aimed at cost-benefit analysis still hits the public choice alternative? An initial point to note is that if the public choice criticism of the benign view of the state assumed by cost-benefit analysis as used in public decision making is correct, then it applies equally well to some of the major critics of cost-benefit analysis. Consider for example Sagoff. Sagoff (1988) distinguishes the preference that individuals have as a consumer and those they have as a citizen. The former express people's private wants, the latter their public values, their judgements about 'what is right or good or appropriate in the circumstances' (Sagoff, 1988, p. 9). Thus, Sagoff, even more than the defender of cost-benefit analysis, assumes that when actors enter the political sphere they take on a quite different personality: that self-interested consumers become citizens concerned with the public good, that politicians express the values of citizens in public law, and that bureaucrats quite neutrally administer that legal expression of public values. While, as will become evident, I have deep misgivings about the assumptions of public choice theory, it is difficult not to have sympathy with their scepticism about any theory that simply assumes that political actors are benign. Indeed, such scepticism need not be associated with free-market economics. Thus, for example, the same scepticism is expressed in a very different political idiom in Marx's early critique of Hegel's benign view of bureaucracy: the bureaucracy does not stand above the egoistic domain of civil society, representing a universal interest; rather it is itself of civil society, the appeal to a universal interest disguising the pursuit of its own interests (Marx, 1843).[7] While it may be false that actors are necessarily and always motivated by narrow concerns with self-advancement, one cannot simply assume in advance that they are not.

However, while there is something right about the public choice critique of state benevolence, other criticisms of cost-benefit analysis, if they are successful, do have purchase against the public choice perspective. In Chapter 3 I have already given some grounds for scepticism about the appeal to property rights that underpins the general case for market solutions to environmental problems. Other criticisms that have particular force in the environmental sphere also apply. For example, the claim that markets cannot properly incorporate concern for the interests of future humans or of existing and future non-humans which necessarily cannot be directly expressed in market choices, still has power (O'Neill, 1993a, ch. 4). Of more general theoretical significance, there are important assumptions that public choice theory shares with cost-benefit analysis, especially concerning the criteria for what constitutes an 'optimal' outcome of either market transactions or government policy. In so far as both are founded in neoclassical theory, they both view as 'optimal', an outcome that most efficiently meets the given wants of the parties concerned. While Austrian economic theory departs from neoclassical standards of optimality, it shares the assumption that any principle of 'optimal' outcomes must itself be purely want-regarding: it takes as given the wants people happen to have and concerns itself with the satisfaction of those wants (Barry, 1990, p. 38). It is this final point I want to examine in some more detail in this chapter.

The choice of want-regarding principles of optimality is standardly justified on one of three grounds:

1 *Meta-ethical*: a subjectivist account of the nature of ethical utterances, according to which they are merely expressions of preferences, entails that principles of public policy be want-regarding.
2 *Welfare*: individual well-being consists in the satisfaction of the wants agents either have, or would have if fully informed, and hence public policy which is concerned with the optimal realisation of welfare must be want-regarding.
3 *Liberal*: public institutions should be neutral between different conceptions of the good, and hence should not be concerned with the cultivation of 'desirable' wants, but only with the satisfaction of whatever wants individuals happen to have. The alternative classical perfectionist view, which conceives of institutions in terms of the pursuit of some particular conception of the good life is, at best, authoritarian, at worst, totalitarian.

At the centre of many critiques of mainstream environmental economics has been a rejection of the purely want-regarding principles of optimality it assumes and also a rejection on the defences that are offered of them (Sagoff, 1988; O'Neill, 1993a, chs. 5–7). Public policy should be concerned not with the satisfaction of given wants, but with the promotion of those institutions which cultivate preferences for what is good. I will not rehearse here in detail the arguments for that position and against those for a purely want-regarding public policy. Very briefly, I believe one should respond to the arguments outlined above as follows:

1 *Against the meta-ethical justification*: a subjectivist account of ethical utterances, even if true, would itself be neutral between first order principles of public policy, which themselves are on that view an expression of preferences; there is nothing in a subjectivist meta-ethics which rules out the 'preference' to cultivate certain wants (O'Neill, 1993a, pp. 64–65).
2 *Against the welfare justification*: the view that preference satisfaction is, of itself, constitutive of well-being cannot be sustained, for desires answer to goods, not goods to desires. I want objects because I believe they are good; I do not believe they are good because I want them (Aristotle, 1908, 1072a 29). Well-being is realised not by satisfying given wants, but by educating our capacities of judgement and our desires, such that we come to prefer what is good (O'Neill, 1993a, pp. 65–82).
3 *Against the liberal justification*: there is no necessary relationship between an authoritarian or totalitarian politics and the defence of institutions in terms of their fostering a particular conception of the good. Any defensible account of the good life needs to recognise the internal plurality of the goods that are constitutive of it. Any defensible account of liberalism needs to engage with

> a defence of substantive values, in particular that of autonomy, and of an account of the institutions that foster them. The best defence of liberalism is one that returns it to its perfectionist roots, that holds with J. S. Mill that 'the first question in respect to any political institutions is, how far they tend to foster in members of the community the various desirable qualities moral and intellectual' (Mill, 1975, p. 29).[8]

I will not offer a detailed elaboration of these responses in this chapter. I have said something in defence of the second response in Chapter 2. I will return to the charge of authoritarianism in Chapter 8. However, there is an assumption that underlies them which does need defence in the present context. The second and third responses depend on a particular view of the relationship between preferences and institutions that public choice theory explicitly denies. They assume that different institutions promote different preferences. The political problem given this assumption is to answer the question Mill asks: what institutions foster desirable preferences? With respect to environmental problems, what institutions nurture concerns for environmental goods? Public choice theory denies both the assumption outlined and the question that is consequent upon it. Preferences are taken to be prior to and explanatory of institutions. Individuals both in market and non-market settings act as rational self-interested agents. The question which, on the public choice account, needs to be answered is this: what institutions should we construct given that individuals are rational self-interested agents, who pursue their own ends both in the marketplace, and in 'nonmarket behavioral settings' (Buchanan, 1972, p. 22). In the next section I defend the Millian assumption and question against its modern public choice opponent.

4.3 Institutional economics: the old and the new

Public choice theory is often presented as a part of a revival of institutional economics. It represents a response, from within the neoclassical tradition, to the neglect of institutional questions in that tradition. Thus, problems concerning the institutional conditions in which markets operate, for example concerning the definition of property rights against which market transactions take place; problems concerning the consequences of the certain institutional forms, such as the unrestrained operation of existing political and bureaucratic institutions; and finally normative problems concerning the specification of optimal institutional arrangements, all become central from the public choice perspective. While this renewed focus on institutions is to be welcomed, for those of us educated outside of this perspective, the neoclassical approach to institutions is still liable to strike one as oddly skewed. There is a clear difference between this new institutional economics and the older institutional economics that traditionally opposed the neoclassicals' institutional myopia (Hodgson, 1988, 1993). I include within the category of old institutional economics not only the American tradition of institutionalism which included Veblen (1919), Commons (1934) and others, but also that economics that took place within a broadly Aristotelian tradition, including the work of both Marx and Polanyi,[9] as well as much of the classical economic tradition.[10]

The central difference between the old institutionalism and the new neoclassical variant is that which I outlined at the end of the previous section. The new institutionalism represents the extension of neoclassical theory into new domains and hence begins with the conventional axioms that define the rational agent within neoclassical theory. In doing so it starts with the assumption that preferences are given. The explanatory problem is to explain the emergence and nature of institutions given that assumption. Reference to institutions appears only in the explanandum, not in the explanans. The central normative problem becomes that of how to fashion institutions given that individuals are egoists.

Old institutionalism differs from the new in that it allows individuals' preferences to be explained by reference to the institutional context in which they operate: references to institutions appear in the explanans, not just in explanandum. Given those assumptions the normative problem becomes that of Mill outlined at the end of the previous section: to determine what institutions should be sustained in order that individuals develop desirable preferences. That assumption underlies the classical political thought of Aristotle. The end of the polis is the good life (Aristotle, 1948, 1280b 38f.), where the good life is characterised in terms of the virtues: hence, the best political association is that which enables every person to act virtuously and live happily (Aristotle, 1948, 1324a 22). Hence also his influential criticism of the market in terms of its encouragement of the desire for the unlimited acquisition of goods and thus the vice of *pleonexia*, the desire to have more than is proper (Aristotle, 1948, Book I, chs. 8–9).[11] The old institutional economics in the wide sense outlined above is the inheritor of this classical tradition.

Old and new institutionalism start from very different explanatory assumptions, and generate distinct normative questions. Which version of institutionalism is to be preferred? The question, in so far as it concerns explanation, is in the end one that has to be answered by reference to the canons of rational inquiry – adequacy to empirical evidence, explanatory power, consistency and so on. However, at present a strong presumption must be made for the old institutionalism. I say this, not only because of the absence of empirical support for many of the standard public choice positions,[12] but also because the new institutionalism has failed in any case to carry out its eliminative project of deleting references to institutions within its explanans. Reference to institutional contexts is smuggled at the level of its assumptions about individuals' conceptions of their interests. Thus while public choice theorists claim to start from preferences that exist prior to institutions, their accounts of the nature of the self-interested preferences of individuals, of their 'utility function', changes according to the institutions they are describing. Within the market it is typically assumed to consist in the acquisition of consumer goods; within the political domain power through the acquisition of votes; within bureaucracy, promotion and advancement in status within the bureaucratic order.

Two points need to be made about these shifting assumptions about the actors' conceptions of self-interest. The first and more basic is that a simple and now familiar observation about the characterisation of action entails that certain interests cannot even be specified outside of a particular institutional context. Consider the politician's interest in the acquisition of votes. Individuals can perform the actions of voting or acquiring vote only when they are embedded in a particular position in an institutional context: it is *qua* citizen that an individual can felicitously vote, and only *qua* candidate that an individual can be elected. Moreover, the action of voting itself depends on a complex set of institutions that embody and are constituted by particular shared understandings. Only within certain institutional settings can the behaviours of marking crosses on papers, the raising of hands and so on be understood as 'voting'. In others, say the raising of the hand in the auction room or the lecture hall, or the marking of crosses against people's names in a classroom, they have different meanings. Hence, an interest in 'acquiring votes' or 'winning an election' is an interest that is only possible within a specific institutional setting. Similar points apply to the interest in 'promotion' and that of 'buying' or 'selling' 'consumer goods': such interests themselves presuppose an institutional context.[13] It is worth adding here, that not only do assumptions about institutional context enter into the descriptions of interests, but also they arrive in the more substantive assumptions about the boundaries between different institutions in the modern world. It is, for example, simply assumed that votes and political office are not the sort of things that, in modern society, can be bought or sold.[14]

The second point about public choice assumptions goes beyond the mere possibility of specifying interests to substantive explanatory problems: that is, in defining individual preferences differently in different contexts, public choice theory implicitly assumes, quite correctly, that different institutional settings foster different conceptions of self-interest. Within the market setting, interests are defined in terms of the acquisition of property rights over objects; within the political domain, power is assumed to be the object of a person's interest; in the realm of bureaucracy, it is identified as the acquisition of status through promotion. The explanatory claims of the older institutionalism enter as unannounced, unnoticed and unwelcome guests into the new institutionalist's assumptions about the 'utility function' of the agent in different contexts. At the explanatory level existing public choice theories have not eliminated reference to institutions from their explanans. Substantive explanatory work has already been done at the level of claims about individuals' conceptions of their interests.

The implicit acknowledgement of the way that different institutions foster different conceptions of an individual's interests has important implications for the public choice theorists claim that they are simply extending the axioms of neoclassical theory concerning the rational self-interested agent into new domains. Two points need to be made here. First, given a full specification of the conceptions of individual self-interest, it is simply not true that one can transfer assumptions about self-interested behaviour in the market to other domains. In other institutional contexts, quite different conceptions of interests are apparent, which can and do conflict with that fostered by the market.

Consider the old conflict between aristocratic and market institutions. In traditional aristocratic societies, honour is institutionally defined as the object of one's interest: to sacrifice one's honour for money would be a sign of vulgarity.[15] This conflict between the bourgeois world of markets and the aristocratic world of honour runs through nineteenth century literature, and any adequate explanation of the cultural shifts in eighteenth and nineteenth century Britain would have to make reference to it. Or to take another example, consider the question of the commercialisation of science. It would be a mistake to see this either, in the fashion of public choice theory, as simply a way of taming 'professionals' who, under the guise of 'scientific values' conspire against the public in the pursuit of the same set of interests they have as 'market actors'; or, as opponents of public choice might have it, as an invasion of a purely 'altruistic' practice (science) by a sphere of egoistic behaviour (markets). It rather involves a shift in individuals' conceptions of their interests. In traditional scientific institutions one's interests were characterised in terms of recognition by peers of a significant contribution to one's discipline, recognition achieved through publication in a peer reviewed journal (Merton, 1968, p. 601; Ravetz, 1973). Commercialised science brings changes in the nature of intellectual property rights such that publication is redefined as an act in conflict with one's interests (O'Neill, 1990, 1998a, ch. 11). Hence the spread of university instructions *not* to publish results, since to do so will be to miss the 'benefit in material terms from the intellectual property you have produced' (from a circular quoted in full in O'Neill, 1992, 1998a, ch. 11). The assumptions about self-interested behaviour in the market cannot be transferred to other institutional contexts. In different roles in different institutions agents have quite distinct conceptions of their interests.

A second point needs, however, to be added to the first. It is far from clear just what assumptions about the economic agent in the market that a public choice theorist is supposed to be taking over into non-market domains. The axioms that define the rational agent in neoclassical theory are a quite minimal attempt to characterise consistency in preferences, a point sometimes noted in its defence. The rational economic agent is assumed to have preferences that are complete, i.e. agents can express preferences over any and all goods; reflexive, i.e. every good is as good as itself; and transitive, i.e. such that if x is preferred to y and y to z then x is preferred to z. The rational economic agent, thus defined, is then assumed to be concerned to maximise the satisfaction of a set of preferences, the 'utility function' in neoclassical jargon, under the constraint of a finite budget. Now, while I believe some of the neoclassical assumptions should be rejected, for example that concerning transitivity (O'Neill, 1998b, 1998c), I do not believe that they should be rejected because they assume an 'egoistic' individual. That individuals are 'self-interested' in the sense that they are concerned to satisfy a consistent set of preferences under budget constraints does not imply that agents are egoists in any strong sense of the term. It all depends what preferences they have: 'The postulate that an agent is characterized by preferences rules out neither the saint nor Genghis Kahn' (Hahn and Hollis, 1979, p. 4).

The rhetorical power of public choice theory depends on its smuggling in through the 'utility function' a particular egoistic characterisation of individuals' preferences. Egoism, in the sense in which it is usually employed, either as a term of derogation, or as a term of political and ethical 'realism', depends on a particular account of the preferences individuals are taken to have. The egoist in the normal sense is an individual who desires only the possession of a narrow set of goods that can be possessed to the exclusion of others: 'the biggest share of money, honors and bodily pleasures' (Aristotle, 1985, 1168b 16) to take Aristotle's list, to which one might add 'political power'. Public choice theory does assume such an egoist with preferences for this narrow range of goods. In doing so it inherits the late-eighteenth-century shift in the language to describe the unlimited acquisitiveness, in which the classical terms *pleonexia*, greed, avarice and love of lucre were replaced by the term 'interest', and hence 'self-interest' was redefined in a narrow fashion (Hirschman, 1977, pp. 31–42). However, in taking for granted this concept of self-interest, it goes beyond the basic formal axioms of neoclassical theory, and implicitly introduces substantive claims about the content of agents' preferences. Moreover, in doing so, its 'realism' becomes quite unrealistic. To an egoist thus conceived the classical response, articulated by Aristotle, forms the proper reply: they have simply misidentified what the goods of life are (Aristotle, 1985, Book ix, ch. 8). Thus, for example, those who are exclusively concerned with the unlimited acquisition of money are improperly called rational economic agents: they are neither 'rational' nor, in the classical sense of the term, 'economic'. The term 'rational economic agent' thus used is a technical euphemism for which the proper description is 'moneygrubber'.[16] Similar points apply to the professional politician driven simply by the desire for political power, the 'politico' or 'hack', or bureaucrat driven by the desire for promotion, the 'careerist'. The derogatory terms employed to describe those individuals express the proper attitude one should have towards them. They are individuals with a hopelessly narrow view of the goods that life has to offer. Moreover, contrary to the 'realism' of public choice, and despite the increasing colonisation of the non-market domains by the market, the recognition that this is so has not entirely disappeared.

It is false to assume that since individuals act as narrowly interested agents in the marketplace, 'the inference should be that they will also act similarly in other and nonmarket behavioral settings' (Buchanan, 1972, p. 22). Individuals are motivated by a variety of ends outside the marketplace: the scientist, by the desire to solve some problem; the ornithologist, by the desire to sustain a habitat in which a variety of birds can be found; the climber, by the desire to climb some new line on a rock face; the musician, by the desire to play a new and technically demanding work; the parent, by the desire to see her child happy and fulfilled; and so on along an endless list. None of this is to deny the existence of egoistic individuals. It is to deny that one can reduce individuals' interests to that narrow set of preferences exhibited in the marketplace and the centres of political power.

The weaknesses of public choice assumptions in this regard are most apparent in their treatment of associations. (The classic public choice treatment of associations is Olson, 1965; see also Olson, 1982.) The term 'association' can refer to a variety of formal and informal societies that are neither direct competitive actors within the market, nor direct competitors for political office, although they can and do have effects on both markets and political outcomes. They include voluntary associations that pursue some particular good – natural history societies, climbing clubs and the like; associations that exist within the economic sphere, but in which actors engage with one another in non-market ways – trade unions, professional associations and so on; organisations that serve some particular interest that is effected by state action – pressure groups, and some charities; the variety of associations that make up the movements and networks of modern civil society; and finally public institutions that are often financed by the state, but are not of the state – universities, schools, hospitals, conservation councils and so on. Such associations form a mixed bag and it is problematic to treat all under the same heading.

Public choice is at its most vulnerable to empirical criticism in its treatment of voluntary associations. Their very existence is a problem given the assumption of a rational actor able to free-ride on the benefits they might bring. As Hirschman (1985) notes of Olson's influential *The Logic of Collective Action* (1965):

> Olson proclaimed the impossibility of collective action for large groups . . . at the precise moment when the Western world was about to be engulfed by an unprecedented wave of public movements, marches, protests, strikes and ideologies.
>
> (Hirschman, 1985, p. 79)[17]

That empirical weakness is a consequence of its assumptions about the nature of 'self-interest'. In public choice theory all associations are treated as 'interest groups', where the term 'interest' is understood in the narrow sense in which it has come to be defined since the eighteenth century.[18] Given that narrow definition of interests pursued by associations, the problem becomes one of how an individual would incur the costs of joining an association rather than free-ride on others. The attempt to reduce all associations to interest groups in that sense is however a mistake. It fails to make proper distinctions between different kinds of associations. Some do exist simply to pursue some narrowly defined interest. However, others exist to pursue some good or 'interest' in the wider sense of the term: consider the wide variety of natural history, conservation and environmental associations. Still others aim both at particular interests and some good: professional associations, even where they are conspiracies against the public, are not *merely* conspiracies as the public choice theorist supposes – they also have an interest in the goods the profession serves, be it medicine, education, philosophy, economics, nature conservation or whatever. Finally, other associations might begin in 'self-interest' narrowly, but develop other interests while in their pursuit, for example in fellowship itself.

Not only does the public choice theorist fail to distinguish between different kinds of association, but also it makes a corresponding failure to distinguish between the different goods or interests an individual may have as a member of an association. A member of an association concerned with the pursuit of some practice, say science, can have two kinds of interests: first, in some achievement internal to the practice itself, in some particular empirical discovery or theoretical development; second, in some external good the association offers, for example, in some form of recognition, in some institutional position, or in an increased salary.[19] The public choice theorist, in implicitly defining interests in a narrow manner, has to treat the first kind of interest as 'really' simply instrumental for the second. However, that is quite implausible. Thus, in many settings, it is difficult to get individuals to fill administrative positions, even where it means promotion, because that would involve sacrificing time on the internal goals that really interest them. Moreover, some of the apparently narrower desires for possessing external goods an institution has to offer have their basis in interests internal to the practice it promotes: a scientist, for example, may desire promotion not out of mere careerism, but because it is a form of recognition of his or her achievements by competent peers.

The weakness of public choice theory when it is applied to associations is that it is precisely in associations that the wide variety of motivations and interests that move individuals is exhibited. In the market individuals do exhibit a preference for the acquisition of consumer goods, in politics, as it exists today, an interest in the achievement of power does predominate, and in bureaucratic organisations, an interest in career advancement is a disposition that is fostered. However, in other associations a preference for a wide variety of goods is apparent. Moreover, as the old institutionalism asserts, such preferences are not givens that are brought to the associations. They are interests that are fostered by them. Indeed, just as an interest in amassing votes is not possible outside a particular institutional context, so many of the interests fostered by associations could not exist without some such institutional context. Thus as Raz (1986) notes, it is only in the context of particular social forms that an interest in 'bird watching' is possible:

> Some comprehensive goals require social institutions for their very possibility. One cannot pursue a legal career except in a society governed by law, one cannot practise medicine except in a society in which such a practice is recognized . . . Activities which do not appear to acquire their character from social forms in fact do so. Bird watching seems to be what any sighted person in the vicinity of birds can do. And so he can, except that that would not make him into a bird watcher. He can be that only in a society where this, or at least some other animal tracking activities, are recognized as leisure activities, and which furthermore share certain attitudes to natural life generally.
>
> (Raz, 1986, pp. 310–311)

Not only are such interests in a wider set of goods distinct from those exhibited in institutional contexts such as the market or politics, but also a commitment to such goods is defined in terms of a refusal to make them commensurable with goods that satisfy the narrow set of interests that define those contexts. Hence, the refusals of individuals to respond to willingness to pay surveys on environmental goods, as we have already discussed. Clearly, on one level, such refusals are not relevant to public choice theory which is not in the business of shadow pricing. However, the refusals are of relevance at another level. As I have already noted in earlier chapters, they uncover widespread and proper convictions about the kinds of things that can be bought and sold. There are commitments that are central to the well-being of agents that are partially constituted by a refusal to put a price on goods. The person who could put a price on friendship, simply could not have friends. They simply do not understand the loyalties that are constitutive of friendship. Moreover they are thereby excluded from much of what is best in human life. Likewise, with respect to other goods that individuals value, including significant places, environments and non-human beings (O'Neill, 1993a, pp. 118–122; see also Raz, 1986, ch. 13). Thus, whatever the truth of Aristotle's comments about 'the many' of the classical Greek world, and one suspects aristocratic prejudices in this regard, the many of the modern world do not exhibit a concern only for that narrow set of goods that is characteristic of the egoist: money, status and power. Individuals engage in a large array of non-market and non-political social associations and practices, and have a correspondingly broader conception of their interests than that ascribed to them by the public choice theorist. The extent to which the behaviour of actors in the market and political world is at all civilised depends on those wider social engagements.

In its saner, more conciliatory and, in the proper sense of the term, 'realistic' moments, public choice theory grants that individuals are not always egoists in the narrow sense, that they are not solely motivated by the desire for money, power and status. The claim is restated in a normative way: that, while it may in fact be false that all persons are egoists, driven by avarice, we need to assume, in the design of good institutions, that they are.[20] This normative claim contains a partial truth. The problem of the vulnerability of institutions to the vicious, and to the careerist, and the lover of lucre and power, are problems that any plausible social and political theory has to take seriously.[21] It does not follow, however, that institutions must thereby be designed around the assumption that *all* persons are thus motivated. The institutions that one would arrive at by that principle are themselves likely to foster the very vices they are designed to check.

One important instance of this point is the familiar case against pure contractarian accounts of good institutions. The contractarian begins from the assumption that institutions be designed around egoists: the only defensible institutions are those which narrowly interested individuals would agree to enter through voluntary contract. The problem with that position is that 'a contract is not sufficient unto itself' (Durkheim, 1964, p. 215): contracts themselves are possible only against the background of non-contractual relations which both build and depend on trust, where trust is an attitude that it is irrational to take given the assumption of universal egoism. The point is one familiar to conservative political thought concerning the ethical presuppositions of the market. It is stated thus by Burke:

> If, as I suspect, modern letters owe more than they are always willing to own to ancient manners, so do other interests which we value full as much as they are worth. Even commerce, and trade, and manufacture, the gods of our oeconomical politicians, are themselves perhaps but creatures; are themselves but effects, which, as first causes, we chose to worship. They certainly grew

> under the same shade in which learning flourished. They too may decay without their natural protecting principles.
>
> (Burke, 1826, p. 155)

Without a background of non-contractual relations, contract itself is impossible.

Universal egoism, in the derogatory sense of the term in which it refers to an interest in the acquisition of possessive goods, is neither a truth about individual behaviour in all institutional settings, nor a sound principle of institutional design. The new institutionalism needs to give way to the old.

4.4 Institutional economics and environmental goods: an agenda

Having attempted to bury public choice theory, as part of a proper funeral oration, I return to two earlier points uttered in its praise. First, it is quite right to insist, against both cost-benefit analysis and many of its recent environmental critics, that one cannot simply assume a benign state inhabited by beneficent state actors, politicians and bureaucrats, who answer either to the preferences of consumers or to the judgements of morally upright voters. Second, the public choice theorist is right to insist that one consider questions about the institutional framework in which environmental decisions take place. However, those questions need to be widened beyond those the public choice theorist allows. The problem is not that of either explaining or designing institutions given universal avarice, but that of examining the ways in which institutions define and foster different conceptions of interests. Individuals' conceptions of their interests needs to be the end point of analysis, not its starting point. Thus explanatory and normative questions posed by new institutionalism should be replaced by those posed by old institutionalism. The central questions to be asked of an institutional environmental economics are these: what institutional frameworks develop a concern for future generations and the non-human world? What frameworks encourage rational argument and debate about environmental matters?

Fortunately, to ask those questions is not to raise issues in a theoretical vacuum. Thus work already exists that starts from those questions in so far as they are concerned with future generations. We have already touched on these in Chapter 3. In an historical paper, Duncan remarks that 'the institutional context in which agronomic decisions are taken should be the *first* thing to be characterised in any general agricultural history' (Duncan, 1992, p. 76, emphasis in the original). Whatever the truth or falsity of his account of the relationship between different forms of lease holding, property ownership, and sustainable agriculture, he is asking the right question. One of the first things that needs to be characterised in an environmental or ecological economics, quite generally, is the institutional conditions of sustainable economic practices: in particular, what institutions foster in individuals a wide conception of their interest that encourages sustainable practices? What institutions undermine such conceptions?[22] In raising those questions, Duncan (1992) is returning to explanatory and normative problems that were central to eighteenth-century debates on commercial society: the civic humanist critics of commercial society were concerned in part precisely with the effects of the commercial mobilisation of property commerce on links between generations (Pocock, 1975, 1985). Neither has that concern ever entirely disappeared. For example, the effects of the mobilisation of land and labour are a central theme in Polanyi's criticism of market society (Polanyi, 1957a, chs. 14–15) to which I will return in Chapter 5. Where institutional economics is less developed is on relationships between institutions and attitudes to the non-human world. Economics needs to move from either concern about shadow pricing or about real pricing, towards some appreciation of the institutional conditions in which individuals begin to appreciate the value of that environment.

Where would that emphasis on the institutional context for environmental goods lead? For reasons outlined in section 4.3, I believe that it will lead analysis away from both market-centred and state-centred approaches to environmental goods and towards an association-centred perspective.[23] The social dimension of environmentalism needs to focus on the question of what associations best develop a concern for environmental goods, and what conditions are required in order that such associations flourish. Even if it turns out that the market and state are institutions we are stuck with, and I retain the possibly utopian hope that they are not, it is the associational background against which they operate that makes the difference as to how far they can and will operate in an ecologically rational manner.[24]

Part II

Time, community, equality

5 Time, narrative and environmental politics

5.1 Time, narrative and separability

How should we value projects and activities over time? Standard cost-benefit analysis offers a clear answer to that question. In cost-benefit appraisal of a project over time, the value of the project (PV) over a time period $t = 0$ and $t = n$ is calculated by the following formula:

$$PV = \sum_{t=0}^{t=n} \{B_t - C_t\}(1+r)^{-t}, \tag{1}$$

where B_t is the benefit at time t,

C_t is the cost at time t,

and r is the discount rate.

If $PV > 0$ then the project is acceptable. Given a choice of projects, the project with the highest total PV is taken to be the best. The use of this formula for project appraisal involves three assumptions:

1. *Discounting the future*: the future should be discounted: future costs and benefits of any proposed option are valued at less than those of the present using some specified discount rate, r, the more distant into the future, the lower the value.
2. *Additive separability*: the values of events at different moments of time are additively separable. Informally, to say they are separable is to say that the value of what happens at some point in time, t_i, is independent of the value at what happens at another point, t_j, and to say that they are additively separable is to say that the total value of a project over a period of time is the sum of these independent values. As Ramsey (1928) puts it in his classic article on saving over generations: 'enjoyments and sacrifices at different times can be calculated independently and added' (Ramsey, 1928, p. 543).[1]
3. *The irrelevance of the past*: the formula assumes some form of consequentialism. What has happened in the past is irrelevant to the value of the project. We always start now, $t = 0$, and consider only the future.

Most of the discussion of the environmental implications of cost-benefit analysis has focused upon the first assumption, and for obvious reasons: it appears to fail tests of equity between us and future generations. Given a discount rate r, a preference measured at £n now is valued in t years' time at £$n/(1 + r)^t$. This has clear implications for our treatment of future generation for it magnifies the value we place on current benefits and diminishes the value we place on future costs: we are justified in shifting costs into the future for our current benefit. However, while I discuss discounting in this chapter, it will not form the central object of concern. My main purpose is rather to question the second and third assumptions and to draw out their implications for our relation to future generations. These assumptions are widely accepted even by those who reject discounting. Ramsey for example accepts both the second and third assumption while rejecting the rationality of discounting (Ramsey, 1928, p. 543). The chapter is in three sections. In section 5.2 I consider the use of assumptions in the appraisal of personal projects for a single person. The use of any such account assumes a particular mistaken view of personal identity over time: the picture of the self and its relation to the future that it presupposes is indefensible. In section 5.3 I draw out the implications of these arguments for the relations of present generations to those of both past and future: if we are to have some sense of community over generations, then the assumptions of additive separability and the irrelevance of the past should be rejected. The assumptions employed to compare costs and benefits in the commercial world are ill suited for evaluation of public projects over time. In section 5.4 I place these particular problems in cost-benefit analysis in the context of more general debates about the disruption of narrative continuity between generations by commercial society.

5.2 The self and its future

One of the central arguments for discounting the future starts from assumptions about individuals' preferences. The argument runs that individuals have pure time preferences – they are impatient: they prefer benefits now to benefits tomorrow and costs tomorrow to costs now simply in virtue of the time they occur; the more distant the future, the less costs and benefits are counted. The argument continues that, since it is the task of social policy to aggregate the preferences of affected parties, time preferences must be incorporated into public policy. Any aggregation of preferences in cost-benefit analysis must reflect these pure time preferences. Now this argument is open to a number of objections. Even if it were to be accepted that social policy should be simply a matter of aggregating given preferences, there is clearly something objectionable about the shift in the argument from intrapersonal preferences to interpersonal preferences. Individuals who have pure time preferences for the present over the future consumption of goods have preferences about their own future preference satisfaction. The aggregation of those preferences does not license any social discount rate in which the preferences of a distinct future population are counted for less than those of existing populations. There is also, however, a prior issue about the rationality of pure time preferences themselves.

There is a standard debate about the rationality of pure time preferences which runs something as follows: the defender of pure time preferences claims we have to respect these preferences. If well-being consists in preference satisfaction, then we must include these preferences in arriving at the policy that maximises well-being. Each individual will have a welfare function that corresponds roughly to the formula given earlier. In considering acts at any point $t = 0$, and looking forward over a probable remaining lifetime to $t = n$, the personal welfare value (*WV*) will calculated by the following formula:

$$WV = \sum_{t=0}^{t=n} \{B_t - C_t\}(1+r)^{-t}. \tag{2}$$

The opponent of pure time preferences holds that a concern with maximising well-being demands that we ignore pure time preferences since to include them would be to fail to maximise the satisfaction of preferences over individuals' lifetimes. Hence Ramsey's remark that the practice of discounting is 'ethically indefensible and arises merely from the weakness of the imagination' (Ramsey, 1928, p. 543). The point is echoed in Pigou's observation that time preferences only show that 'our telescopic faculty is defective' (Pigou, 1952, p. 25). One should take a temporally neutral perspective on the satisfaction of preferences across an individual's life time. Pure time preferences are irrational. The welfare value will be represented rather by a simple sum:

$$WV = \sum_{t=0}^{t=n} \{B_t - C_t\}. \tag{3}$$

Both of these perspectives share the assumptions that the value of events at different moments of time are additively separable and that the past is irrelevant. Are these assumptions defensible? Consider first the assumption of additive separability, the claim that the value of what happens at some point in time, t_i, is independent of the value at what happens at another point, t_j, and the total value over a period of time is the sum of these independent values. Is this assumption defensible? Consider the following examples that I have used elsewhere (O'Neill, 1993a, p. 53):

A. A newly married couple, couple A, go on a two-week honeymoon. The holiday begins disastrously: they each discover much in the other which they had not noticed before, and they dislike what they find. The first four days are spent in an almighty row. However, while they argue continuously over the next eight days, they begin to resolve their differences and come to a deeper appreciation of each other. Over the last two days of the holiday they are much happier and both feel that they have realised a relationship that is better than that which they had before their argument. The holiday ends happily. Sadly, on their return journey, the plane that carries them explodes and they die.

B. A newly married couple, couple B, go on honeymoon. The first twelve days proceed wonderfully. On the thirteenth day their relationship deteriorates badly as each begins to notice and dislike in the other a character trait which they had not noticed before, at the same time realising that the other had a quite mistaken view of themselves. On the last day of the holiday they have a terrible row, and sit at opposite ends of the plane on the return journey. They both die in an explosion on the plane.

Which lives go better? Or, to stay with the language of consumer choice, given a visitation on the day before the holiday begins by an angel who presents you with a choice between the two lives, which would you choose? (The visitation and the choice itself will be instantly forgotten, so can be ignored.) Given the assumption of additive separability, then whether or not one discounts the future, the answer looks as if it should be B. Thus given a temporally neutral perspective, we have a sum given by (3) above. We are to tot up the days of dissatisfaction – in the case of A, the first twelve days, in the case of B the last two days – and the days of satisfaction – in the case of A the last two days and in the case of B the first twelve days – and the total value is the sum of these. Given rough equality in dissatisfactions from arguments and satisfactions from enjoyable times together, holiday B comes out as the better choice. If one adds a discount rate and employs the formula given in (2) the value of B over A is simply magnified: since the dissatisfactions in B happen later, they count for less at the point of choice on the visitation by my angel. So given additive separability and a few assumptions about the costs over benefits scores for each day of the holiday, B looks like the best consumer value.

However, many individuals (although not all as I have discovered in offering the choice to some of my more hedonistically inclined students), given a visitation from my angel, would choose holiday A. They characterise the story of holiday A as a happier one than that of holiday B. They do so because the temporal order of events is neither irrelevant nor is it the case that it is always better to have goods sooner rather than later. What counts in favour of holiday A is the narrative order of events, and crucial to that order is the way in which that story ends.

However, to assume the additive separability of value through time is to assume that the order of events does not matter, *as such*, to the value of the total. It matters only through anticipations and memories themselves experiences to be valued at a moment of time. The total value of a project over a period of time is simply taken to be the sum of the value of each moment taken in separation. The sequence of events is irrelevant to its total value. The value of each event in a moment of time can be ascertained separately, in isolation from what occurs before and after. Now clearly there are theories of value which entail such separability. Thus, as I noted above, some of my more hedonistically inclined students are inclined to choose holiday B, and from within a hedonistic perspective this makes sense. If one assumes, with Pigou, a subjective state account of welfare, that 'the elements of welfare are states of consciousness' (Pigou, 1952, p. 107), be these states of pleasure or satisfaction, then one can treat the value of events in time as separable for whether or not the event was pleasurable or painful can be ascertained independently of what happened before or after. Thus while it might be the case that the anticipation or recollection of an event causes pain or pleasure, its characterisation as painful or pleasurable at the time at which it occurred cannot be altered. From with a hedonistic perspective, the order of events as such cannot matter. However, this simply gives us I believe one more reason for rejecting a hedonistic account of value. For the narrative order of events does matter, and that it does entails that the assumption of separability has to be rejected.

There are characterisations that we make of events in a person's life that can be made only in the context of a larger narrative order. Thus the way we characterise the days in the honeymoon stories depends on their place in a larger narrative frame. In holiday A, the argument at the start of the holiday is not simply a 'cost' – a moment of pain or desire dissatisfaction. Rather, taken in context, it might be taken to be a 'turning point' in the relationship, one which clarifies the relationship and lays the foundation for the ensuing happiness. Within the context of the individuals' entire lives, it has another significance. For that reason one can also talk of the earlier event having been 'redeemed' by the later reconciliation to which it gave rise. Likewise, the moments of happiness in holiday B are not 'benefits' – feelings of satisfaction. Rather, within the context of the whole story, they are moments of illusion, when each person has a false view of the other, an illusion shattered by the final argument. Had their lives continued, the argument also may have become something else, but the ill fortune of untimely deaths robs the participants of such a future. Whether moments of pain and pleasure are goods or evils depends on their context of a life as a whole. They do not come ready-tagged as such.

To make these claims is to deny the separability of the value of moments of time. An evaluation of what happens at one moment in time cannot be made independently of what happens at others. A painful argument can be characterised positively only as a 'turning point' in a person's life given that the events that follow are of a certain kind. A reconciliation counts as a 'redemption' of events only given its particular relation to those past events from which it grew. For that reason, when we are living through events, painful or pleasurable, it is often difficult to ascertain what their final value will be. It depends upon their place in the larger story of our lives and whether this turns out well or ill depends upon the way that story unfolds and ends: the genre to which the story belongs depends upon the way it concludes. Given that this is the case temporal separability, and more specifically additive separability, fails. One cannot treat the total value of a person's life over a period of time by totting up the values of individual moments taken in isolation.

Because temporal separability fails in this particular way so also does the irrelevance of the past. Given the place of narrative orders in evaluation, it makes sense to make decisions that refer not only to the future, but also to the past. The argument here is not that standardly invoked against consequentialism – that there are acts in the past such as promising that lay upon us current obligations. Rather, it is that we can act now in ways that make intelligible or redeem our past. A farmer might quite rationally shun the enticements of a good life with his children in a city and live out his days alone and in hardship on a hill farm, in virtue of the fact that the farm represent his life work: to leave and to let it go to ruin would be to render his life pointless. A student with a terminal illness may continue to an examination at the end of the course because it is the culmination of her past studies, even if it looks on a strictly consequentialist reasoning to make no sense at all.

The principle of separability of moments of time presuppose a particular picture of the self and its continuity over time. Given separability the relation between my self and my future takes on a strangely impersonal aspect. A strong sense of identity over time that is given by narrative unity or continuity is lacking. Life is treated as a series of discrete episodes, and our relation to the future begins to look like a relation to other persons. The only argument is how much I should care for the future selves that follow me. For the defender of time preferences at any moment t_0 O'Neill cares for O'Neill$_0$ who exists now more than O'Neill$_1$ who exists at t_1 who is a physically related relative of O'Neill$_0$ who in turn is cared for more than the more distant relative O'Neill$_2$ and so on as time stretches out into the future. The critic of pure time preferences assumes the same picture but is an economic maximiser who demands equal consideration for future preferences on the basis of an oddly impersonal perspective. A life is still a series of discrete acts of consumption, but the maximiser asks for any future O'Neill, O'Neill$_n$ to be given the same consideration as the current O'Neill, O'Neill$_0$. The life still lacks internal coherence. The only relation between the various O'Neills is one of physical continuity. The relation of a person to the future becomes one of self to others.

This picture of temporal neutrality has indeed been exploited by utilitarians. Sidgwick raises the following objection to the Egoist understood in the uncommon-sensical sense as one who refuses the Utilitarian principle of universal benevolence:

> If the Utilitarian has to answer the question, 'Why should I sacrifice my own happiness for the greater happiness of another?', it must surely be admissible to ask the Egoist, 'Why should I sacrifice a present pleasure for one in the future? Why should I concern myself about my own future feelings any more than about the feelings of other persons?'
>
> (Sidgwick, 1907, p. 418)

The thought, developed in detail by Nagel (1970) and Parfit (1984, part two), is that a self-interest theory of rationality, according to which each person aims at the world which makes his or her life go as well as possible, is vitiated by the fact that it is agent-relative, i.e. it gives reasons that are reasons only for the agent, but not time-relative, i.e. it aims to maximise well-being across time. The claim is that the rejection of impersonality and the acceptance of temporal neutrality cannot be happily combined. My relation to my future selves is not in principle different from my relation to others. Any future 'self' is an other like any other.

This account of the self to its future needs to be rejected. The picture of identity over time it assumes is in error. What is absent is precisely a view of human lives as ones that have a narrative structure – as stories of physical and mental growth and development, of decline, of the success and failure of projects and relationships. It fails to allow any place for narrative order in our identity and correspondingly in the evaluation of a moment or period in our lives. The future of one's own life does matter for an appraisal of current life in a way that the current or future lives of strangers need not. What the future is like matters to how well we can say our present life is going. Our present life is part of a larger narrative and the shape of that whole life matters. In particular, the way matters turn out matters. Our lives are not or should not be a series of disconnected events between disconnected selves such that at any moment we can say now whether our lives are going well or badly. Our own future matters to us because it determines what appraisal we can give to the present. Hence the truth in Solon's dictum that we can call no man happy until dead (Aristotle, 1985, Book I, chs. 10–13). The problem with Solon's dictum is only the point of death is too soon to make an evaluation: a person's death is not the end of the narratives of which they are part.

5.3 Between past and future generations: community, environment and public decisions

The assumptions of the separability in the value of different moments of time and the irrelevance of the past may fail in the valuation of how well an individual's life can be said to go. It does not follow immediately that they fail also for public social decisions about projects over time. Hence it does not follow immediately that there is any problem with cost-benefit analysis as a decision procedure for assessing the value of projects over time. Do the arguments in section 5.2 about intrapersonal relations of past, present and future have any implications for interpersonal relations between individuals in different, albeit overlapping, generations?

One initial reason for thinking it may have such significance is the way in which considerations about the past are peculiarly relevant to deliberations about the environment. A feature of deliberation about environmental value is that history matters and constrains our decisions as to what kind of future is appropriate. This is true of the conservation of natural objects. We value natural objects, forests, lakes, mountains and ecosystems specifically for the particular history they embody. The very ascription of 'naturalness' to them depends upon the specific history we can tell of them. The block on 'faking' nature (Elliot, 1982, 1997) lies not just in the origin of natural objects but also in the history that takes us from their origin. However, the significance of history is still more evident in the problems that are typical of nature conservation in the European context. Many have arisen as a consequence of too abrupt changes in the use of the land – the loss of meadowland and pastures with the use of nitrogen and intensive farming techniques, radical shifts in patterns of grazing, the disappearance of hedgerow and copses to make way for efficient use of new machinery, the drainage of old wetland rescued from the sea. Most of these nature conservation problems are concerned with flora and fauna that flourish in particular sites that are the result of a specific history of human pastoral and agricultural activity, not with sites that existed prior to human intervention. Nor is this a typically 'old world' phenomena. Many of the conservation issues in the new world have the same form: indeed their presentation as problems of 'wilderness preservation' are misleading if one means by a wilderness a landscape or habitat that is undisturbed by human activity. Rather, many new world 'wildernesses' in Africa, Australia and America are the old world pastoral landscapes shaped by those who had previously lived in those places, and whose activities are rendered invisible and their memory lost. That this is so has been a source of many of the practical problems in 'managing wilderness' for the 'wilderness' as discovered by Europeans was a landscape that depended upon pastoral activities such as burning.[2]

This relevance of the past is evident also in the conservation of the embodiments of the work of past generations that are a part of the landscapes of the old world: stone walls, terraces, thingmounts, old irrigation systems and so on. And at the local level past matters in the value we put upon place (Clifford and King, 1993). The value of specific locations is often a consequence of the way that the life of a community is embodied within it. Historical ties of community have a material dimension in both human and natural landscapes within which a community dwells. The city is an environment that has suffered its own dislocations: hence the complaint 'I no longer know the place'. Place is also valued on a more individual level – the local familiar walk, stream, pond or landscape that invokes a very specific past of one's childhood, parents, friends, or whatever. Hence the strong reactions roused by external changes that transform those places. In all these cases, the past is relevant to our current evaluations. It is because a particular place embodies a particular past that they cannot be substituted for by new artificial replacements. The history matters. Place like nature cannot be faked. We aim to preserve an ancient meadowland, a thingmount, an ancient stonewall and not modern reproductions. As such the past does constrain our current decisions.

The constraint that the past places upon us is best understood as one founded in the significance of the narrative orders objects and places can have for us. We enter worlds that are rich with past histories, the narratives of lives and communities from which our own lives take significance. The problem is, or should be construed as, the problem of how best to continue the narrative; and the question we should ask is: what would make the most appropriate trajectory from what has gone before? The value in these situations which we should be seeking to uphold lies in the way that the constituent items and the places which they occupy are intertwined with and embody the life-history of the community of which they form a part. This is the perspective which lies behind the following attempt to characterise the objectives of conservation: 'conservation is . . . about preserving the future *as a realisation of the potential of the past* . . . [it] is about negotiating the transition from past to future in such a way as to secure the transfer of maximum significance' (Holland and Rawles, 1994, p. 37, emphasis in the original). It is because it is about the negotiation of a narrative order between past and future, that it is not simply about 'preserving the past'. Indeed once the significance of the role of a narrative order is understood, we can understand more clearly just what is wrong with that approach. One major problem with the heritage industry is the way it often attempts to freeze historical development. A place then ceases to have a continuing story to tell. The object becomes a mere spectacle, a museum piece, taken outside of any common history.

Public decisions, no less than private choices, do not start from year zero. They are, or should be, expressive of a particular understanding of relation of past and future and the communities to which we belong. They are contested for that reason. For the same objects and places often embody quite different narratives which sometimes point to different and conflicting trajectories between which we must adjudicate. There are different histories to which we have to be true – and there are histories that, when they are unearthed, change our perceptions of the nature of a place and what it embodies. For example the empty hills of highland Scotland embody not just a wild beauty but also the absence of those who were driven from their homes in the clearances. Their memory must also be respected. The argument over the fate of the 100 foot high statue to the Duke of Sutherland on Beinn Bhraggie Hill near Golspie is about which history we choose to acknowledge. Between 1814 and 1819, the 'Black Duke', as he was known, played a leading role in evicting the local people from their homes (Craig, 1990). We should perhaps support its removal not just on aesthetic grounds but also for what it represents to the local people, some of whom are descended from those who were driven out. For the same reasons, the now dilapidated cottages which their ancestors left behind should perhaps remain. More controversially, such histories form a proper part of the debate as to how far the highlands of Scotland should remain unpopulated.

Finally, it needs to be noted that in discussing the transition from past to future we are not mere conduits through which history flows. The future matters not just to the past but to us – for the characterisation of how well our own lives go depends upon what happens after we have gone. There are a variety of projects in which we engage and relationships in which we are involved which are such that how well our lives can be said to go can depend on what happens to the projects and relationships that occur beyond our lifetime. Hence my remark at the end of section 5.2, that Solon is wrong: at death it is too soon to say how well a person's life has gone. Consider what it is to engage in the sciences and the arts. The status of a piece of scientific work depends on its relation to both a particular past and a particular future. In relation to the past, a piece of scientific work makes sense only within a particular history of problems and theories to which it makes a contribution. Its success or failure depends on its capacity to solve existing problems where others fail. However, it also depends on a projected relation to the future in terms of its capacity to solve not just existing problems, but also problems unenvisaged by its author, and in its fruitfulness in creating new problems to be solved and new avenues of research beyond anything a scientist may have considered. Correspondingly, that there exist future scientists educated in a discipline and able to continue work within it matters for current scientific activity.

Likewise the greatness of many works of art lies in their continuing to illuminate human problems and predicaments in contexts quite foreign to that in which they were originally constructed. Many of the aesthetic qualities of a work of art may only become apparent in virtue of its relation to future works. Hence the truth in T. S. Eliot's comment: 'the past should be altered by the present as much as the present is directed by the past' (Eliot, 1951, p. 15). For this reason, it is of significance for us now that there be future generations able to appreciate the arts and contribute to them. Similar points apply to more everyday activities, raising and educating children, tilling soils, creating towns and cities in which the future can be happily lived. In all such activities, the future matters to us now. This is apt to sound paradoxical, especially when stated in terms of our being 'harmed after our deaths' (O'Neill, 1993b): the phrase conjures up implication of backward causation and is resisted for this reason. However, this worry misses the point, for no such implication is involved. The point is rather that how we characterise the value of an event depends upon a larger narrative frame in which it occurs and that frame itself depends upon a particular future. It is because in our lives both public and private, the value of different events does not answer to the assumptions of separability.

It is just this point that highlights the problem with market-based decision-making procedures like cost-benefit analysis. As I noted at the outset of this chapter, cost-benefit analysis assumes additive separability of moments of time and the irrelevance of the past. As such it disrupts proper deliberative reflection of our relation of past to future. It involves the introduction of particular assumptions about the public appraisal of projects that are central to institutions in commercial society. In the world of costs and benefits in a competitive commercial society the past is irrelevant, for the survival always depends upon future profitability. The past and long-term future count only to the extent they may be represented in consumer preferences. Moreover, specifically commercial values, the prices individuals are willing to place on objects and events at specific moments of time, are additively separable over time.

Cost-benefit analysis takes these commercial norms into the world of public choices. However, whatever the appropriateness they have for the economic sphere, and I return to this more general consideration in section 5.4, they are inappropriate for public choice. They disrupt the social narratives that make possible community over time. Hence, decision procedures like cost-benefit analysis are apt to foster forms of development and planning that are ahistorical and placeless. The justification of the road that runs through copse, woodland and ancient meadowland to save a few minutes on travel time, expresses an attitude to time as a series of separable moments that come as cost or benefit, not as the narrative order of a life. Such results are not accidental. Given the assumptions of additive separability and the irrelevance of the past, one can expect decisions that fail to appreciate the significance of places that bind communities over time. In public choices about the environment we express the kinds of relations to past and future we want as a society: the very assumptions of cost-benefit analysis rule out proper expression of those values.

5.4 A brief social history of time

The disruption of a narratory continuity between past and future by commercial society is a theme that has its own history. As I noted in Chapter 3, the disruption of historical narrative by market norms lay at heart of debates about land and commerce in Britain in the seventeenth and eighteenth centuries. The civic humanist criticism of commercial society was founded on claims concerning the relationship between the civic virtues and ownership in landed property. Commercial society, by mobilising land, undermined that link between generations. While that debate has been less evident in much recent political and economics theory, it never entirely disappeared. This relation between the market and the mobilisation of land has been echoed by later socialist critics of commercial society. It is not just the mobilisation of landed property by the market that undermines intergenerational identity, but also the mobilisation of labour. Specific ties to a particular locality and place, to a stable extended community within a locality, and commitments to a particular craft and profession are inimical to and undermined by the workings of a market society. The theme is particularly evident in Polanyi. Both land and labour are fictitious commodities, objects treated as if they were produced for sale upon the market. As fictitious commodities they are bought and sold in real markets and the consequence is the disruption of social ties of place (Polanyi, 1957a, p. 73). Workers in a market society must be prepared to shift location and occupation if they are to achieve the market price for their labour. The ties of 'a human community to the locality where it is', ties of place, are undermined. To have a tie to a place is to have a tie with an environment which reveals a particular past history (Polanyi, 1957a, p. 184). It is to recognise the skills embodied in buildings, landscapes and objects that form our environment and to have a sense of continuity with those whose skills are thus made public. To disrupt ties to place is not merely to remove persons from a particular spatial location, but also to divorce them from relations to previous generations and a sense of continuity with the future.

Historical continuity is similarly disrupted by the mobilisation of labour across occupations. The advocacy of the mobility of labour was central to the defence of the market economy by its early defenders like Adam Smith whose work was aimed against guilds and the practice of lengthy apprenticeship as a barrier to the movement of labour. Again a major response was that ties across generations are weakened by the disappearance of continuity in craft and work. The relation of craftsman and apprentice is undermined, and with it the sense in which success in craft work was tied to past and future. The view is echoed in Weil's comment:

> A corporation, or guild, was a link between the dead, the living and those yet unborn, within the framework of a certain specified occupation. There is nothing today which can be said to exist, however remotely, for the carrying out of such a function.
>
> (Weil, 1952, p. 96)

The mobilisation of labour by the market, like the mobilisation of land, has undermined a sense of community across generations.

These critical accounts of the effects of commercial society on our relations to past and future are I believe substantially correct. However, to make them is to raise a problem rather than a solution. There are good reasons to reject a possible corollary that we return to stable ownership of the land and limited mobility in labour. The particular ties of pre-modern societies were often oppressive, and the dissolution of old identities a liberation from personal servitude and narrow horizons. Moreover, even if it were desirable to limit mobility, in modern conditions it would not be practicable without excessive coercion. The problem of obligations to future generations is a social and political problem concerning the economic, social and cultural conditions for the existence and expression of narrative identity that extends across generations. At the heart of that issue is the problem which has been the focus of much social and political theory for the past two centuries – that of developing forms of community which no longer leave the individual stripped of particular ties to others, but which are compatible with the sense of individual autonomy and the richness of needs that the disintegration of older identities also produced. The recent debate between communitarians and liberals is a variation on a well-established theme and itself restates the problem rather than offering a new solution.[3]

One recent turn in social theory which will clearly not do as a solution to that problem is that of postmodernism in which a radical aesthetic model of autonomy is asserted at the expense of identity and narrative. The postmodernists owe a large debt to the situationist discussions of the problems of narrative and time in the modern commercial society. The situationists, and Debord in particular, were acute critics of the way historical time had been disrupted in commercial society founded upon commodity production (Debord, 1977, chs. V–VI). The criticism, somewhat different from that of the earlier critics of commercial society, plays on the way that time is sold and consumed: blocks of time are sold, life is presented as a 'sequence of falsely individualised moments' (Debord, 1977, para. 149); individuals' lives are fragmented such that the story of development is lost; a life of frozen at youth is promised. 'The spectator's consciousness . . . no longer experiences its life as a passage toward self realizations and toward death' (Debord, 1977, para. 160).[4] There is I think a great deal to be said in defence of this view of the way modern market economies disrupt the narrative continuity of individual lives. Where I am less convinced of the situationist analysis is in the ways in which they believe that this disruption of narrative can be overcome. Their criticism of the disruption of narrative is married with an aesthetic account of individual autonomy as playful self-creation which is empty of content: offered 'a playful model of irreversible time of individuals and groups' or a 'program of total realization, within the context of time, of communism which suppresses "all that exists independently of individuals"' (Debord, 1977, para. 163), I remain not just unconvinced but unclear what it is supposed to be like as a possible lived experience. The demands remind me of a comment a friend made many years ago that the trouble with the situationists is that they wanted the adverts to be true.

The trouble with postmodernists is that they think the adverts are true. Indeed the distinction between appearance and reality is largely rejected. Much in postmodernist theory is situationism without the social criticism. Where the situationists remained critical of the disruption of individual narrative lives in commercial society, the postmodernists see already existent in consumer culture the possibility of a 'playful attitude' to self and narrative. The result is the celebration of the disruption of narrative in modern commercial society. The kind of fragmented individual life of merely physically related selves that is central to economists' accounts of time reappears as the fragmented self who is able to play with her or his identity. However, the aesthetic account of autonomy, of a chooser who can put on and take off identity at will, is mistaken. To able to play with your identity is to have none. To have an identity is to be the possible subject of a coherent life narrative (O'Neill, 1998a, ch. 6).

The celebration of the disruption of narrative continuity in individual lives is reproduced by postmodernists at the social level. Postmodern theory has been characterised in terms of the loss of the credibility of 'grand narratives' (Lyotard, 1984, p. 37). Now there is a truth contained in this scepticism about grand narratives. What it gets right is that, at any moment, there is always more than one story to be told. As we have seen, in the environmental sphere a place is always the meeting points of a multiplicity of narratives, some of which are silenced. Consider again the hidden human histories that are lost in the name of 'wilderness' to which I will return in Chapter 7. These need to be resurrected and doing so alters our understanding of what is at stake in the landscape. Genealogy, the favourite critical tool of postmodern thought, works by unmasking troubled histories. It is premised on the importance of history in evaluation.

However, postmodern incredulity about grand narratives not only is about uncovering the pluralism of stories that can be truly told, but also involves a rejection of the very possibility of a coherent narrative in either individual or collective lives. This has its basis in a scepticism about certain particular 'grand' narratives – what Lyotard calls the narratives of emancipation and speculation – a scepticism founded largely in disillusionment with the Marxist narratives which writers like Lyotard once accepted. Now whether or not one is right to reject these particular narratives, and I believe they have more going for them than is often assumed, their rejection does not entail the rejection of the very possibility of social narratives.[5] What the temporal dimensions of environmental problems show is both how appeal to narrative continuity at both individual and social levels continues to have some kind of evaluative force and correspondingly a need to have institutional forms that allow us to regain a social sense of narrative continuity over time that commercial society has disrupted. The critical force that the concept of narrative played in discussions of commercial society needs to rescued from its loss in postmodern analysis.[6]

6 Sustainability

Ethics, politics and the environment

'Sustainability' and 'sustainable development' have become key phrases of the politics of the environment. The concept of sustainability began its life in environmental debates in particular contexts such as agriculture, forestry and fishing in which it is typically used to characterise the long-term durability of a specific set of practices. A fishing policy that leads to the depletion of stocks in particular fishing grounds, or agricultural practices that lead to irreversible losses of topsoil in a particular area are said to be unsustainable. Similar uses of the concept are employed with respect to human impacts on specific habitats and ecosystems. However, in much recent work in economics and political theory, and more generally in policy documents the concept of sustainability has taken on a more general usage, especially when used to describe what is termed sustainable economic development. The publication of the Brundtland Report was particularly influential on this development. The 'Brundtland' formulation of the concept of sustainable development, taken from the 1987 report of the World Commission on Environment and Development chaired by Gro Harlem Brundtland, reads as follows: 'Sustainable development is development that meets the needs of the present without compromising the ability of future generations to meet their own needs' (World Commission on Environment and Development, 1987, p. 43). The report goes on to say that in interpreting 'needs' here, overriding priority should be given to the essential needs of the world's poor.

Since the Brundtland Report the terms 'sustainability' and 'sustainable development' appear regularly not only in the language of environmentalists but also in national and international policy documents and the rhetoric of mission statements in businesses. This widespread use of the term has been a source of understandable suspicion. This is strengthened when the term 'sustainable' is used to modify nouns like 'growth' and 'development' in ways which appear to disguise the possibility of conflicts between continued capital expansion and other environmental goals. Among environmentalists there is a worry that the terms are simply being used as green rhetoric for continuing economic growth as usual. It is not hard to find spokespersons for business announcing that they too believe in sustainability, for after all they have to sustain their business and customer base. Worries about the concept of sustainability are at the same time apparent among critics of environmentalism who claim that the concept is muddled and serves to place claims of 'the environment' above the more pressing needs of the existing poor.

The very elasticity of these uses of the concept of sustainability raises questions about what it is supposed to mean: the sustainability of what, for whom, and why? In this chapter I consider these prior questions in more detail

6.1 Sustainability, justice and equality

The concept of sustainability raises problems that are distributional in nature. The Brundtland definition includes, quite properly, issues of distribution within generations, and I will return to that dimension later. However, the focus of the sustainability debate has been about the distribution of benefits and harms over time. Many of the harmful environmental consequences of current actions will fall upon those who follow, while many of the benefits will come to some, although by no means all, of those who live now. Environmental concerns bring to the fore problems of the proper distribution of goods and harms over generations. The standard presentation of that issue in the literature is in terms of intergenerational justice or equity. That framing of the question highlights a connection between arguments about sustainability and a number of recent arguments in political theory concerning the nature and defensibility of equality. It raises the first of debates around different answers to a question that Sen in particular has been responsible for bringing to the fore in discussions of equality: 'Equality of what?' (Sen, 1980). There are a variety of different answers to that question in the literature: that we should aim at equality of welfare understood as preference satisfaction, equality of resources (Dworkin, 1981), equality of liberty and in the distribution of primary goods (Rawls, 1972), equality of opportunities of welfare (Arneson, 1989), equality of access to advantage (G. A. Cohen, 1989), equality of capabilities to functionings (Sen, 1980, 1997), equality in goods that fulfil objective interests and needs. The debate is clearly of significance for arguments around sustainability. If sustainability is about equity in distribution over generations, then it raises the same question as to what it is we are supposed to be distributing equally. The question 'Equality of what?' is directly related to the question 'Sustainability of what?' And while the issue is not always raised in this form, one way of understanding the debates between the various conceptions of sustainability is that they differ on what we are supposed to be distributing: for example, whether we should be aiming at sustaining levels of well-being understood as preference satisfaction (Pearce, 1993, p. 48), options or opportunities for welfare satisfaction (Barry, 1999), resources, environmental capacities (Jacobs, 1991 chs 7 and 8, 1995), capacities to meet needs or objective interests (World Commission on Environment and Development, 1987).

The question, 'Equality of what?', is clearly related to a second general question, 'Why equality?' Derek Parfit (1997) has drawn a series of distinctions between different answers to that question which have been influential on subsequent discussion. The first is that between telic and deontic egalitarianism: telic egalitarianism is the view that we should promote equality because it is a good outcome in itself and correspondingly that what is wrong with inequality is that it is a bad state of affairs as such; deontic egalitarianism is the claim that we should aim at equality because to do so is to perform the right or just action and correspondingly that what is objectionable about inequality is that it involves wrong-doing or injustice? A second distinction that Parfit (1997) draws is between egalitarianism of either kind and what he calls 'the priority view', the view that benefiting people matters more, the worse off they are. We should all other things being equal choose the action that produces the greatest improvement in the well-being of the worst off. The view is a version of what Temkin has described as 'extended humanitarianism' (Temkin, 1993, ch. 9). Extended humanitarians, it is sometimes claimed, are not strictly speaking egalitarians, although they may defend similar redistributive ends: redistributions are defended in virtue of the specific claims of the less well off and their greater urgency compared to those of the better off, not because equality as such is a good.

How exhaustive is Parfit's typology? Consider for example the following characterisation of the ideal of equality by George Orwell:

> Up here in Aragon one was among tens of thousands of people, mainly though not entirely of working-class origin, all living at the same level and mingling on terms of equality. In theory it was perfect equality, and even in practice it was not far from it. There is a sense in which it would be true to say that one was experiencing a foretaste of Socialism, by which I mean that the prevailing mental atmosphere was that of socialism. Many of the normal motives of civilized life – snobbishness, money grubbing, fear of the boss, etc. – had simply ceased to exist. The ordinary class-division of society had disappeared to an extent that was almost unthinkable in the money-tainted air of England; there was no one there except the peasants and ourselves, and no one owned anyone else as his master.
>
> (Orwell, 1966, pp. 101–102)

The ideal of equality here is tied to a wider set of virtues than that of justice. Indeed justice in a narrow sense is not mentioned. The appeal of equality here is that it is a condition for a certain kind of community and human character.

Consider one objection to the priority view or extended humanitarianism as a complete account of the value equality. If we can marginally improve the condition of the worst off by large increases in the condition of the best off, then the extended humanitarian will not have any objection as such to the policy. The distribution of goods as such does not matter. However, the egalitarian might want to suggest that this misses the source of the value of equality. Inequality may be objectionable even where the condition of the least well off is improved. Why? One answer is that justice is not done. The distributions still violate fairness. However, another would appeal to the kind of community that would involve. Inequality is a constitutive condition of vices such as dependence, humiliation, snobbery, servility and sycophancy. Central to the history of egalitarian thought is the argument that equality is a condition for a certain kind of community and human character. The ideal of equality is tied to the creation of a community in which certain forms of power, exploitation and humiliation are eliminated and solidarity and fellowship fostered (Tawney, 1964; Miller, 1997; O'Neill, 2001a). It might even be that an egalitarian community requires departures from strict justice, but that it is none the worse for that: justice is one virtue among others and we are willing to allow it to be subordinate to others – for example, generosity or mercy. To defend equality by appeal to the relationships and virtues of character it fosters is not necessarily to take equality to have merely instrumental value in the sense of being an external means to a distinct end: equality and the mutual recognition of equality are partially constitutive of many of the virtues and relationships to which appeal is made (Miller, 1997; Norman, 1997; O'Neill, 2001a).

This latter approach to equality has important implications for the question 'Equality of what?' The answers to the questions I outlined earlier tend to focus on some external objects or states of affairs to which individuals might have different levels of access. While distribution in this sense is important, and nothing I say in the following should be taken to deny that fact, it tends to miss other significant dimensions of equality. Equality is concerned with the nature of the social relationships between individuals. It is concerned with issues of power, of relationships of superordination and subordination, and with issues of recognition, of who has standing in a community and what activities and attributes are acknowledged as having worth. The poor do not just lack the means to satisfy their needs, important as this is. They also tend to be powerless and socially invisible, their activities not given proper recognition.

What implications do these debates between different arguments for equality have for debates about sustainability? Consider sustainability in so far as it concerns intergenerational distribution. To the extent that sustainability concerns intergenerational equity, a number of distinct possible answers present themselves. Given telic egalitarianism that extends over time, it would be, in one respect at least, good in itself to realise equality over generations, even where these include distant generations whom we do not know and to whom we are unrelated. The deontic egalitarian would defend intergenerational equity on the grounds that we should do what is just to future generations, while to give them less is to commit an injustice. The egalitarian who appeals to equality as constitutive of a certain kind of community would appear to need to appeal to equality as a condition of a certain kind of community across generations. Finally, the extended humanitarian will argue that we should distribute goods to benefit most those who are worse off. Priority is to be given to the poor. The different positions entail different justifications and ends in intergenerational justice. How far any of these positions are adequate is an issue to which I return below.

6.2 Intergenerational justice and sustainability

How should we understand our obligations to future generations? One obvious approach to that problem is to attempt to extend notions of obligations of justice from those that apply to current existing persons and groups to those that apply over time. However, doing so has raised a number of problems that have been at the centre of recent debates on our obligations to future generations. Here I will look at two of those problems: the absence of circumstances of justice and the apparently inegalitarian nature of positive obligations to future generations.

The circumstances of justice

The first set of problems with extending conceptions of justice and ethical obligation that apply to current generations to future generations is that concerned with the absence of some of the circumstances of justice and ethical obligations on certain views of those obligations. Two views of our ethical ties are particularly vulnerable to this problem. The first view is that offered by certain version of contractarian theories of justice and obligation, according to which the point of moral rules or more narrowly the rules of justice is to serve as a means by which individuals of limited altruism can realise their long-term interests in conditions where they are roughly equal in power and vulnerability. Such theories have problems in extending considerations of justice or morality to future generations, for given that there is an inequality of power and vulnerability (we can harm them, they cannot harm us) the circumstances of justice or obligation are absent. The second view is that of those accounts of justice that link particular obligations to others to the existence of community – the view that many obligations exist only in conditions in which we belong to the same community. Given that future generations either cannot or may not belong to a community with ourselves, what kind of responsibilities do we have for them?

There are a number of responses one might make given these problems. First, one might accept the ethical theories, and follow through the implications that we have only limited obligations to future generations.[1] Second, one might reject those conceptions of justice or ethical obligations since they fail to offer an account of our obligations to future generations. Third, one might attempt to defend such conceptions by modifying them so as to include intergenerational obligations – for example by introducing the notion of an intergenerational contract – or by denying the supplementary claims, for example that we cannot be harmed by future generations or that future generations and ourselves cannot belong to the same community. My own view is that the second response is owed to contractual theories of ethical duties, but, with some qualification which I will discuss below, the third response is owed to theories that relate justice to community.

In Chapter 5 I defended the claim that the future matters to us now. The argument runs roughly as follows: many of the projects we engage in, scientific, artistic, familial, political and everyday working activities, depend for their point and their potential success on a future beyond us. It matters to us that future generations do belong to a community with ourselves – that they are capable, for example, of appreciating works of science and art, the goods of the non-human environment, and the worth of the embodiments of human skills, and are capable of contributing to these goods. This is an obligation not only to future generations, but also to ourselves that we do not undermine our own achievements by rendering impossible our own success, and to those of the past, so that their achievements continue to be both appreciated and extended. On this view our obligations should not be understood as obligations to strangers but to members a transgenerational community of which we are potentially a part. In my previous formulations of this claim I have sometimes stated the point in terms of our being 'harmed after our deaths'

(O'Neill, 1993b). As I noted in Chapter 5, thus stated the point is one that is apt to sound paradoxical, since the phrase conjures up implication of backward causation and is resisted for this reason. For that reason I would now be more reluctant to use that phrase. However, the central point relies on no such assumption. The central point is that how we characterise the value of events in our lives depends upon a larger narrative frame in which they occur and that frame itself depends upon a particular future. Hence, the future matters to us now.

However, while I have defended this approach, it does have a limitation. It does not deal well with obligations to temporally distant strangers. The problem is one captured by one of my favourite pieces of graffiti: written on the perimeter fence of a construction site building a prestigious new cultural centre in Melbourne, it read 'and this too shall pass'. And it shall. There may well come a time when our cultural and scientific achievements are lost, reduced to electronic parchments that none can read, and our cities buried beneath debris. There may come to exist people who have followed historical disruptions, for whom we are at best distant curiosities and with whom they have no significant cultural continuity. It would be hubristic to expect this might not happen. If this happens, it would still be wrong to leave those people a bequest of nuclear and industrial wastes that we know will do them harm. Indeed, there is a strong case for the view that some such bequests will be a particular problem just when continuity in the scientific and cultural conditions for the skills necessary for dealing with such problems have been lost.[2] We have obligations to future generations of a negative kind that are not adequately captured by an approach that relies entirely on ties of community. It may be that this requires an appeal to more impartial accounts of justice.[3] However, while this may be true, it is still the case that, while it does not offer a complete account of inter-generational obligations, narrative and transgenerational community matters more than is often supposed. Indeed I think there are reasons to think that in the sustainability debate they may do more work than is often supposed. This emerges if we consider a second problem of extending standard accounts of justice to relations between generations, those concerning duties to improve the conditions of those who follow us.

Chronological unfairness

The second problem concerns the apparently inegalitarian nature of many positive obligations that are taken to apply over generations, with what Herzen called the chronological unfairness of human development.[4] The point is one that has been pushed by Beckerman (1999; cf. Beckerman and Pasek, 2001, chs. 4–5) who has questioned how far the language of 'sustainable development' captures any principle of egalitarian justice: sustainable development is often couched in terms not of *equality* of welfare over generations, but of *improving* the welfare or quality of life of future generations. While negative obligations that concern avoiding harms to those in the future do often lend themselves to being couched in terms of justice or equity, these positive duties raise problems. With regard to negative obligations there is something prima facie unjust about our engaging in projects in which the benefits fall on ourselves while the harms fall on those who follow us. There is also a negative sense in which harms and suffering visited on one generation for the benefit of future generations can also be said to be unjust: Stalin's policy of sacrificing one generation for all those that follow lends itself to criticism on those grounds (Deutscher, 1966, p. 296). One also might understand how such obligations might be grounded in considerations of equality outlined above. However, when it comes to positive obligations notions of intergenerational justice and equity do not always appear to be the right language in which to articulate our concerns.

The problem here is that normally our obligations to future generations are understood in general terms as obligations to leave them a state of the world *as least as good* as the one we inherited, but if possible better. I take it this need not mean merely that we leave them 'materially better off' – at some point there may be limits to how far material improvements are possible although where those are is a matter of debate – but better off in terms of social, cultural and environmental goods they inherit. To quote a passage of Marx we discussed in Chapter 3:

> a whole society, a nation, or even all simultaneously existing societies taken together, are not the owners of the globe. They are only its possessors, its usufructuaries, and, like *bone patres familias*, they must hand it down to succeeding generations in an improved condition.
>
> (Marx, 1972, p. 776)

While this principle, that we should leave not just as good, but better to future generations is sometimes taken to invoke a principle of justice,[5] it is far from clear that it should be. It is certainly not a principle of equality in any of the senses we have outlined them. It asserts that it would be better if future generations would be better off than the present or at least be better endowed with opportunities and resources than current generations. It entails an unequal distribution of goods over time, a rising curve in which the later are better off than the earlier. Nor, given such a curve, could it possibly be defended on the humanitarian ground of giving priority to the worst off. The point is noted by Rawls (1972) who for this reason does not apply the difference principle to savings over generations. If one did, then there should be no savings at all by any generation, g_n, which would make the following generations better off than g_n, since then g_n would be the worst off generation relative to subsequent generations and by the difference principle inequalities are justifiable only if they improve the situation of the worst off (Rawls, 1972, p. 291).

If the injunction to improve the condition of those who follow us is not a principle of equality or extended humanitarianism, is it still a principle of justice of another kind? I cannot see how it can be. It does appear to fit better Herzen's characterisation of human development as a form of chronological unfairness. For similar reasons Kant found the duty to improve the condition of humankind 'puzzling':

> [N]ature does not seem to have been concerned with seeing that man should live agreeably, but with seeing that he should work his way onwards to make

> himself by his own conduct worthy of life and well-being. What remains disconcerting about all this is firstly, that the earlier generations seem to perform their laborious tasks only for the sake of the later ones, so as to prepare them for a further state from which they can raise still higher the structure intended by nature; and secondly, that only the later generations will in fact have the good fortune to inhabit the building on which a whole series of their forefathers (admittedly, without conscious intention) had worked without themselves being able to share in the happiness they were preparing.
>
> (Kant, 1784a, p. 44)

From the point of view of justice there are reasons for Kant's puzzlement. The principle, 'population A has an obligation to make population B better off than they are', not only doesn't look like a principle of justice, but also, applied between groups within a generation, looks very like an ideology of exploitation. There are contexts in which population B has power over A that it would be a version of that: for example, something very like that principle underlay some traditional justifications for rendering women's interests subservient to those of men.

What justifications are there for the principle we ought to leave future generations better off than ourselves? One response, defended by Beckerman (1999), is that there is no good justification for the principle of sustainable development. While I think he may be right to reject the claim that anything in justice, equity or extended humanitarianism requires the positive obligations involved in the principle of sustainable development, it is not clear to me that the principle fails for that reason. There are other sources of justification that could govern our dealings over time. A possible justification of the principle that could do the work is the appeal to utilitarianism, that we should maximise well-being. One policy that might be expected to maximise well-being over time would be for each generation to improve the condition of its successor, without making excessive sacrifices itself – that it enjoyed at least some of its inheritance bequeathed it by previous generations. However, the appeal to utilitarianism has clear problems. First, some paths of sustainable development may not maximise total utility. Second, utilitarianism as an aggregative principle for maximising total utility is any case notorious for permitting apparent injustice. Objectionably exploitative versions of the principle could also survive the utilitarian test. Some account is required how what in other circumstances would be objectionably exploitative is not so here. I want to suggest that the historical community-based view of our relations to the future defended in the Chapter 5 does offer a better account than the alternatives of our positive obligations and why they are compatible with justice.

Consider the nature of the obligations involved. There are a series of relationships, practices and projects in which success is defined in the terms of the history of a community and where to leave those who follow in a condition that is better than one finds it is a constitutive condition of success. Thus consider Marx's formulation of the principle. While there is a certain irony in his choice of familial relations to introduce the principle, it is well chosen. It is with relations of parent to children that the sentiment is most explicit: good parents do want their children to have opportunities and goods in their lives that are, if possible, better than in their own. And while there are versions of this demand that have pernicious effects, particularly where children, at the cost of their own autonomy, become the surrogate means to the realisation of their parents' ambitions and projects, the principle is here quite justifiable. It captures one of the virtues that define what it is to be a good parent. This is true of the virtues of persons in other human practices. There are a number of projects which make sense only in terms of improving options to others within a history of a community. It is true of the sciences, arts, agriculture, the building of cities and communities. In a straightforward sense it is part of what it is to do well in the sciences, for example, that one leaves one's discipline in a better state than when one arrived. The worth of one's work is defined in terms of its place in the history of a discipline, the problems it solves before and the contributions it makes to what follows. It defines what it is to have been a successful scientist. Newton not only stood on the shoulders of giants, but also was himself stood upon. There is a quite clear sense in which the relations of present to past and future to present are exploitative but in a neutral rather than pejorative sense.[6] We exploit our predecessors and posterity will exploit us. But to be thus exploited in this sense is a measure of success. For that reason while the obligation to improve the conditions of those that follow are not duties of justice, they are compatible with justice. Being exploited here is not normally an injustice on us, unless our role is forced and lived to the cost of other components of our well-being – as were those of Stalin's policies. Exploitation becomes objectionable where the relative vulnerability and powerlessness of one party is a condition of being used by others. While there may be cases in which this is true – consider again some ideologies of motherhood that denied women any other projects in their lives – for the cases outlined exploitation in this sense is absent. The principle that we improve the condition of those who follow us does make sense and is compatible with justice given an understanding of the roles of narrative and community in understanding the overall assessments we make of our lives.

6.3 Sustainability: weak and strong

I noted at the outset of this chapter that the issue of sustainability raised the questions: The sustaining of what? For whom? Why? In the mainstream economic literature on sustainability the answer to those question runs roughly thus:

1. What is to be sustained? A certain level of human welfare, where in standard welfare economics this is understood as preference satisfaction.[7]
2. For whom it is to be sustained? Present and future generations of humans.
3. Why? Either to maximise welfare over time, or to meet the demands of distributional justice between generations.

In its basic sense sustainable development is defined as economic and social development that maintains a certain minimum level of human welfare. Pearce for example defines sustainability as follows:

> 'Sustainability' therefore implies something about maintaining the level of human well-being so that it might improve but at least never declines (or, not more than temporarily, anyway). Interpreted this way, sustainable development becomes equivalent to some requirement that well-being does not decline through time.
>
> (Pearce, 1993, p. 48)

Sustainability on this definition raises the question as to what is required so that a certain level of human welfare be maintained over time. In much of the economic literature on sustainability the answer to that question is offered in the language of 'capital'. The maintenance of a certain minimum level of human welfare over generations requires each generation to leave its successor a stock of capital assets no less than it receives. It requires that capital should be constant, or at any rate not decline, over time. If, however, sustainability is defined in these terms it might be asked just what work the concept is doing that could not be done without it (Beckerman, 1994). What special issues do specifically environmental problems raise?

The response of defenders of the concept of sustainability in economics to the objection that it has nothing new or useful to add to the debate on the distribution of goods over time, is to insist on the importance of particular states of the natural world for the welfare of future generations. The point is still stated in terms of 'capital', specifically in terms of the concept of 'natural capital'. A distinction is thus drawn between natural and human-made capital: human-made capital includes physical items such as machines, roads and buildings and 'human capital' such as knowledge, skills and capabilities; natural capital includes organic and inorganic resources construed in the widest possible sense to cover not just physical items but also for example genetic information, biodiversity, ecosystemic functions and waste assimilation capacity. The distinction between the two forms of capital is taken to generate two possible versions of the sustainability requirement, each with variations:

> (i) that overall capital – the total comprising both natural and human-made capital – should not decline, or
> (ii) that natural capital in particular should not decline.
>
> (Pearce et al. 1989, p. 34)

This distinction is expressed in terms of that between 'weak' and 'strong' sustainability.

The debates over weak and strong sustainability turn on the degree to which 'natural capital' and 'human-made capital' can be substituted for each other. Proponents of 'weak' sustainability are often taken by critics to affirm that natural capital and human-made capital are indefinitely or even infinitely substitutable. Proponents of strong sustainability, on the other hand, hold that because there are limits to which natural capital can be replaced or substituted by human-made capital, sustainability requires that we maintain the level of natural capital, or at any rate that we maintain natural capital at or above the level which is judged to be 'critical'. We are thus offered the following definitions:

> Weak sustainability assumes that manmade and natural capital are basically substitutes.
>
> (Daly, 1995, p. 570)

> Strong sustainability assumes that manmade and natural capital are basically complements.
>
> (Daly, 1995, p. 570)

The question becomes an empirical matter of how far 'natural' and 'human' capital can be substituted for one another and how far they are 'complements'.

In fact, as Alan Holland has noted, the differences between the positions maybe exaggerated, with both sides in the debate setting up straw opponents (Holland, 1997). Accounts of weak sustainability are more cautious than defenders of strong sustainability often suggest. Beckerman notes that weak sustainability 'allows for substitutability between different forms of natural capital and manmade capital, *provided that*, on balance, there is no decline in welfare' (Beckerman, 1994, p. 195, emphasis added). And even Solow, who is often taken to be the most notable proponent of weak sustainability, asserts 'the world can, in effect, get along without natural resources *if* it is very easy to substitute other factors for natural resources' (Solow, 1974, p. 11, emphasis added). Neither asserts unconditionally that natural and manmade capital are substitutable. Indeed, it is hard to find any economist who makes the claim that 'natural' and 'human' capital are infinitely substitutable. The response of the neoclassical economist is to claim that substitution between natural and human capital at the margins is at issue, not total substitutability of one for the other (Beckerman, 2000; cf. Arrow, 1997, p. 579). Correspondingly an important and under-explored issue concerns just where and how far the concept of marginal changes applies in the environmental sphere.[8] On the other hand critics of strong sustainability like Beckerman often set up something of an easy target. Beckerman ascribes to defenders of strong sustainability the view that no species should be allowed to go extinct, nor any non-renewable resource exploited no matter what the cost in poverty for current generations, which allows him to criticise the concepts as both morally repugnant and impractical. Defenders of strong sustainability rightly reject that version of their position. Given that neither side actually holds the position to which their opponents ascribe to them, what actually differentiates the positions? Is there a problem in principle with the standard neoclassical approaches?

While the statements of writers like Beckerman and Solow are much more careful than their critics often suggest, the more general concerns about the extent to which neoclassical economic theory encourages an excessively optimistic view of the possibility of substitutability are well founded. The assumption of ubiquitous substitutability of goods is built into the intellectual apparatus of the neoclassical framework.[9] Consider for example the smooth continuous indifference curves of economic textbooks built on the assumption that for a marginal loss in one good there is a marginal gain in another which leaves the agent at the same level of preference satisfaction. What are the sources of this assumption of substitutability of goods? One possible source is the picture of the human agent that underpins neoclassical economics. The agent is a rational agent, who maximises preference satisfaction given formal constraints of the structure of his or her preferences – that they are transitive, complete, continuous – and budgetary constraints on meeting those preferences. The account of the human agent is one that abstracts from the way that actual agents are physically and socially embedded in the world. And it is when we turn to agents as physical and social beings that the limits of substitutability become clear. As physical beings actual human agents have physical needs which require satisfaction from distinct physical sources that are not substitutable. A person with malnutrition requires specific objects of nutrition: more entertainment or better housing will be no substitute. Neither, if they suffer a vitamin deficiency, will more proteins do. Correspondingly particular natural and agricultural environments matter because they bring physical goods satisfying bodily needs for which no substitutes are possible. Water, air, nutrition and the causal preconditions for these do not have substitutes. As socially embedded beings, particular relations to particular persons, to kin and friend, and particular places matter in ways which do not allow for substitution. No platonic cousin can substitute for a particular brother. As I noted in Chapter 5, particular places have value in virtue of embodying the particular histories of a community, and that particular history blocks their substitution by places that are physically similar. Hence, for reasons I develop in more detail below, no restoration project can substitute for a particular place in which the particular history of a particular community is embodied. It is then in the particular features of human beings as physical and social agents that blocks on substitutability become apparent. If one considers only abstract rational agents those blocks will not be apparent. It is as physically embodied and socially embedded agents that the limits of substitution come to the fore.

Now I think it is open for a neoclassical theorist to simply accept these points. In the idealisations of economic theory we can assume goods that are substitutable at the margin and draw continuous indifference curves. That does not commit the theorist to assume that all goods are so substitutable any more than a physicist's idealisation of frictionless surfaces commits him to the assumption that actual surfaces are frictionless. A distinction needs to be drawn between theoretical idealisations and actual cases and so long as it is recognised no problems need emerge. The real limits of substitution can be acknowledged if economic theory is suitably modest about its own status. This plea for a modest economic theory has I think something to be said for it – all theory depends on idealisations. However, it is not a modesty that is always observed. Where idealisations are treated as actual descriptions the result is unsatisfactory in theory and pernicious in policy practice.[10] Moreover, even granted these points about idealisation, there is still something in the nature of the standard neoclassical welfare economics which encourages excessive optimism about the substitutability of goods. Idealisation is not the only source of the problem.

A second source of this optimism about the substitutability of goods in standard neoclassical welfare theory is the preference theory of welfare it assumes. On a preference satisfaction account of well-being, well-being consists in the satisfaction of preferences, the stronger the preference the great the improvement in well-being. In Chapter 1 I have already noted how this account leads to a misplaced optimism about the substitutability of goods. Given a preference satisfaction account of well-being, goods are substitutable for each other provided total preference satisfaction levels remain unchanged.[11] This assumption lies at the basis of the indifference curves of neoclassical economics. An agent is taken to remain indifferent between a marginal loss of one good and a marginal gain in another. The agent is indifferent between the two bundles of goods. Preference satisfaction levels remain unchanged. Goods are substitutable for each other at the margin. As I noted in Chapter 1, if we reject this preference satisfaction account of well-being for an objective account then there is no reason to assume that goods will be substitutable in this way.[12]

Given an objective list account of well-being which includes an irreducible plurality of components to what makes life go well there is no reason to assume that goods will be ubiquitously substitutable for each other. Consider Nussbaum's (2000) list of the central human functional capabilities, where these are broadly categorised under different headings: life; bodily health; bodily integrity; senses, imagination and thought; emotions; practical reason; affiliation; other species; play; control over one's political and material environment (Nussbaum, 2000, pp. 78–80). Each heading defines a space of capabilities to function. Each space can be itself internally plural, including an irreducible variety of capabilities. Each might also be open to being realised in different way. For example, there are different ways in which concern for the world of nature can be realised and expressed, a point to which I will return in section 8.3. However, given an irreducible plurality to different human functional capabilities, the assumption of ubiquitous substitutability between different goods with respect to welfare should be rejected. For a marginal loss of a good in one space, such as bodily health, it is not necessarily the case that there is a marginal gain in other, say practical reason, that leaves the person's well-being unchanged. A loss in one dimension may be properly addressed only by a gain in the same dimension. As I put it above, a person who suffers from malnutrition requires specific objects of nutrition: more entertainment or better housing will be no substitute. Or to return to our example in Chapter 1, the loss of the place in which the life of a community has been lived through the construction of a dam does not have substitutes in gains in other dimensions. A loss in the dimension of affiliation cannot be compensated for by a gain in other dimensions. Hence the good sense in the everyday observation that there is no substitute for good health, for good friends, for particular places and environments. To say this is not to dispute the existence of hard choices in private and public life in which some will suffer. Nor is it to deny that in those conditions something is owed to those who suffer by way of recompense. What is mistaken is the claim that there is a level of compensation, a sum of money, that would maintain their level of welfare. The assumption that there is is based on a false theory of welfare.

These points apply across generations. Sustaining or improving welfare requires the realisation of a variety of different capabilities. What we need to pass on to future generations is a bundle of goods that can maintain welfare across the different dimensions of human functioning – we need to able to pass down the conditions for life and good health, for social affiliation, for the development of capacities for practical reason, for engaging with the wider natural world. To do that requires that the goods we pass on are goods that are disaggregated and which cannot be substituted for each other. Practical reason requires formal and informal institutions for education. More and better entertainment on television will not do. Sustaining affiliation requires sustaining the cultural and physical conditions for community, including particular environments that are constitutive of communities. An increase in the number and quality of consumer goods will not be a substitute. Maintaining the capacity to appreciate the natural world and to care for other species requires us to sustain particular environments. Again increases in entertainment will not substitute for that particular good. There are a variety of different human capabilities that require distinct and non-substitutable goods to realise them.[13] It is not simply that 'natural' and 'human' 'capital' are not substitutable for each other or are in some sense complementary goods. It is rather that a various environmental goods are not substitutable by other goods because they meet quite distinct dimensions of human well-being. Moreover, there are particular human-made environments which form places that are constitutive of particular communities that also lack substitutes. I return to the point below.

6.4 Nature without capital

Both sides of the debate between strong and weak conceptions of sustainability employ the metaphor of 'natural capital'. The use of the metaphor involves an 'asset'-based approach to those questions. To refer to the natural world as 'natural *capital*' is to construe it in a particular way as a 'stock' of assets that forms 'a source of benefit streams' for humans. On one view, which is often taken to form the basis of 'weak sustainability', this stock is given a monetary value through use of valuation techniques discussed in earlier chapters. However, even if one rejects that picture, as I believe one should for reasons outlined earlier, to retain the metaphor of 'natural capital' is to conceive of the natural world as a stock of assets which give a stream of benefits for humans. The assets are understood as capacities to 'provide humankind with the services of resource provision, waste assimilation, amenity and life support' (Jacobs, 1995, p. 582). The question of the acceptability of substitution turns then on whether one stock of assets can be replaced by another and still maintain the same capacities. Now this dimension of the relations of humans to the natural world is clearly an important one. What proponents of strong sustainability are right to note is that there are real physical limits to the capacities of nature to deliver those services and to human capacities to find human substitutes for all of these natural services. However, there are reasons to be sceptical about how far the language of 'capital' is the right metaphor to capture this dimension of our relation to the natural world. It uses as a general metaphor a very specific set of relations of humans to each other and nature that has emerged in commercial societies of the past 300 years or so. The fact that the metaphor of natural capital lends itself to monetisation is neither accidental nor as surprising as Jacobs (1995) suggests. The term 'capital' in everyday parlance is used in a commercial context.

However, even given that it is possible to divorce the term 'natural capital' from this specific commercial context, the metaphor is ill suited to capture other relations of humans to the natural world. What is involved in sustaining 'nature' or sustaining 'the value of nature'? If one starts with the approach suggested by the metaphor of 'natural capital' then the result is something as follows. A list of goods is offered that correspond to different valued features of our natural environment – an itemisation of different species and habitats – and sustaining nature or nature's value consists in maintaining or enhancing the numbers on the list. We have something like a score card, with valued kinds of objects and properties, valued goods, a score for the significance of each and we are to maintain and where possible increase the total score, the total amount of value.

Consider, for example, one of English Nature's position statements on sustainable development. It defines 'environmental sustainability' thus: '*environmental sustainability* . . . means maintaining the environment's natural qualities and characteristics and its capacity to fulfil its full range of activities, including the maintenance of biodiversity' (English Nature, 1993, emphasis in the original). Having thus defined the concept, the position statement continues thus:

> Those aspects of native biodiversity which cannot be readily replaced, such as ancient woodlands, we call *critical natural capital*. Others, which should not be allowed, in total, to fall below minimum levels, but which could be created elsewhere within the same Natural Area, such as other types of woodland, we refer to as *constant natural assets*.
>
> (English Nature, 1993, emphases in the original)

Sustainability is taken to involve the protection of 'critical natural capital' – that part of the natural environment which cannot be 'readily replaced' – and a set of 'constant natural assets' that do have possible substitutes through recreation and translocation. In using the terms critical natural capital the document is at the stronger end of the sustainability debate. The use of the term 'natural' in this context does have a strained meaning which is acknowledged. In the context of the United Kingdom, given that almost all habitats have a history of human use, the distinction between 'cultivated' and 'natural' is hard to make. It becomes even harder when the maintenance of 'constant natural assets' is conceived in terms of a deliberate policy of human made habitat creation or translocation.

What are the criteria of adequate 'substitutability' or 'replaceability'? Capital, both natural and human-made, is conceived of as a bundle of assets. We have a list of valued items, habitat types, woodlands, heathlands, lowland grasslands, peatlands and species assemblages. We maintain our natural capital if, for any loss of these, we can recreate another with the same assemblages. The promise of the approach is its flexibility. If a road potentially runs through some rare habitat type, say a meadowland, or an airport runway is to occupy woodland that contains some rare orchid, we can allow the development to take place *provided* we can recreate or translocate the habitat. The issue becomes one of the technical feasibility of replacement: on this turns the distinction between 'critical natural capital' and 'constant natural assets'. It also offers the possibility of allowing damage to some landscape, on the condition that afterwards the ecological scene can be restored, or even enhanced.

What are the limits of replaceability? The *UK Biodiversity Action Plan* (HM Government, 1994) states the problem of replaceability in a way that appears to allow a large sphere of critical natural capital:

> While some simple habitats, particularly those populated by mobile species which are good colonisers, have some potential for re-creation, the majority of terrestrial habitats are the result of complex events spanning many centuries which defy re-creation over decades. Therefore, the priority must be to sustain the best examples of native habitats where they have survived rather than attempting to move or re-create them elsewhere when their present location is inconvenient because of immediate development proposals.
>
> (HM Government, 1994, para. 3.96, p. 43)

Elsewhere the plan adds:

> [H]abitats are of great significance in their own right, having developed initially through colonisation of the UK from the rest of north-west Europe after the last glaciation and then subsequently under the direction and influence of traditional human land management activities. The results of these long historical processes are not reproducible over short time scales, and indeed like individual species themselves, are a product of evolution combined with chance events which cannot be re-run the same a second time.
>
> (HM Government, 1994, para. 3.95, p. 43)

The issue of what belongs to critical natural capital and what does not turns on the technical limits of what can be replaced. The 'results' of 'long historical processes' cannot be reproduced readily. As a report for English Nature puts it: 'the environmental conditions that moulded them cannot be technically or financially re-created within acceptable time scales' (Gillespie and Shepherd, 1995: 14). Hence the references to ancient woodlands as a particular habitat that is an example of critical natural capital. Were it possible to re-create within acceptable time scales, then we could shift such habitats into the box of constant natural assets. *This Common Inheritance* (HM Government, 1990) comes up with the time scale of 25 years – a human generation. We thus have a picture of two kinds of habitat, that which it is technically relatively easy to replace, the pond, secondary woodland, secondary heathland, meadowland, which can be shifted around to fit development and those which take longer to recreate which are to be permanent features of the UK landscape. The approach leaves open the possible technical development of translocation or re-creation skills allowing a shift in habitats from the status of critical natural capital to that of constant natural assets.

Is that approach to sustainability of nature satisfactory? The role of time and history I outlined in Chapter 5 point I think to deeper problems with an approach to environmental sustainability which understands it purely in terms of maintaining some stock of natural items. Time and history do not enter into problems of environmental policy just as technical constraints on the possibility of recreating certain landscapes with certain physical properties. Particular habitats are valued precisely because they embody a certain history and processes. The history and processes of their creation matter, not just the physical attributes they display. The temporal 'technical constraints' of the UK biodiversity plan are in fact a source of the very value of habitats that could not in principle be overcome. We value an ancient woodland in virtue of the history of human and natural processes that together went into making it: it embodies the work of human generations and the chance colonisation of species and has value because of the processes that made it what it is. No reproduction could have the same value, because its history is wrong.

In deliberation about environmental value history and process matters and constrains our decisions as to what kind of future is appropriate. We value forests, lakes, mountains, wetlands and other habitats specifically for the particular history they embody. Geological features have histories with no human component, while landscapes often involve the interplay of human use and natural processes. Most nature conservation problems are concerned with flora and fauna that flourish in particular sites that are the result of a specific history of human pastoral and agricultural activity, not with sites that existed prior to human intervention. The past is also embodied in the work of past generations that are a part of the landscapes of the old world: hedgerows, stone walls, terraces, buildings and so on.

The natural world, landscapes humanised and worked through pastoral and agricultural activity, the built environment, all take their value from the specific histories they contain. We do not enter into or live within 'natural capital'. Our lived worlds are rarely natural and are not capital. We live in places – habitats if one likes – that are rich with past histories, the narratives of lives and communities from which our own lives take significance. We need to take the narrative and temporal dimensions seriously. We do not live in capital or stocks or bundles of assets. We live in places that are significant in a variety of different ways for different communities and individuals. And the natural world in which humans have entered and will one day leave is that, a natural world with its own history: it is not 'capital'. If we are destroying marshes and forests, we are destroying places and habitats, not just a stock of assets. Environments, plural, are not merely bundles of resources. They are where human lives go on, places to which humans have a lived relation of work, struggle, wonder and dwelling. The failure of the metaphor of natural capital to capture the significance of the historical dimensions of environmental values is symptomatic of a failure to capture adequately the different dimensions of the relations of humans to their environments.

To thus criticise the concept of 'natural capital' is not necessarily to reject the concept of sustainability. The language of sustainability does have rhetorical power

– in the main because in all its technical uses it retains something of its everyday meaning, where 'to sustain' is 'to maintain the life of something'. Reference to sustaining land, children, forests, future generations, communities has power since it calls upon a metaphor of keeping going something that has a life – in either a literal or a metaphorical sense. For this reason the language of sustainability does have some real force in specific contexts. I am not suggesting that the language should be abandoned. However, the use of the metaphor of capital is ultimately unhelpful as a way of filling out this idea. It is a metaphor that is drawn from a historically specific set of market relations between humans in their dealings with nature. It reduces our relations to nature to one – that of resource – and loses sight of the other relations we have to natural world. Indeed, it would rather be better to hold onto the everyday sense of the term. Something that has a life has a history and potentiality to develop. To sustain the life of a community or land is not to preserve it, or to freeze it but to allow it to change and develop from a particular past into a future. It is also to allow there are conditions in which mourning death and loss is appropriate rather than attempting to keep things going as end in itself. Sustainability needs to be placed in the context of political and ethical debates about the kinds of communities over time to which we understand ourselves to belong.

6.5 Environmental justice within generations

As I noted at the start of Chapter 5, the central focus of recent work on environmental philosophy has been on issues of justice and ethical obligation to non-humans and future generations. In one way this is understandable, since these are the issues highlighted by specifically environmental problems. However, it also has some unfortunate consequences. The focus has tended to reinforce the tendency to concentrate on the distribution of goods and harms, at the expense of questions about the distribution of power and recognition. At the same time, it has seen a lack of acknowledgement of the way that environmental problems raise issues of equity between current generations of humans which are just as pressing in a practical sense as those that have dominated ethical debate in environmental philosophy (Martinez-Alier and O'Connor, 1996; Guha and Martinez-Alier, 1997; Martinez-Alier, 2002). Environmental decisions take place in the context of existing inequalities in property and power; they have consequences for the subsequent distribution of damages and goods across different groups, often falling hardest on the poorest. The siting of toxic waste dumps and incinerators, of open cast mines, roads, runways and power stations, all adversely affect the health and quality of life of those communities who will have to live with them. They are often felt by those without the power, voice and wealth to resist their imposition. Hence they are distributed unevenly across class, gender and ethnic groups. The effects of deforestation in the developing world will often fall differentially on women. The creation of national parks in the developing world has brought benefits to visiting tourists, local elites and possibly some wildlife: it has often involved the displacement of communities that have lived and worked in those areas from their traditional pastures and hunting grounds into marginal lands. The introduction of large commercial fishing fleets into traditional fishing waters has lead to the decline of poorer fishing communities. The construction of dams involves the displacement of communities and the loss of homes and ways of life. The arguments over biodiversity have a strong distributional dimension. The attempt to create new systems of intellectual property rights on genetic resources are can bring benefit of large commercial firms to the detriment of those developing world communities in which much of the genetic diversity is situated and whose agricultural practices have sustained.

Despite the universality of some environmental risks, it is still the poor that suffer disproportionately from environmental harms. In particular, in the context of a market system in which goods and harms are valued by people's willingness to pay to receive or forgo them, environmental harms will fall disproportionally upon the poor. That this is the case also highlights the existence of resistance to environmental harms from the poor. Hence Martinez-Alier notes against the view that modern environmentalism is a 'post-materialist' movement, that there are significant forms of environmentalisms of the poor that are responses to those material environmental problems (Guha and Martinez-Alier 1997; Martinez-Alier, 2002; cf. Leff, 2000).

The problem is not just about distribution of environmental resources, services and burdens. Indeed, as I noted above, the focus on environmental distribution within current generations of humans highlights dimensions of inequality that have tended to be overlooked in environmental discussion. The question 'Equality of what?' has answers that refer not only to external objects or states of affairs to which individuals might have different levels of access, but also to the nature of the social relationships between individuals and groups: with issues of power between groups, of domination and subordination; and with issues of recognition concerning which actors and activities are socially acknowledged as having worth. With respect to both future generations and the non-human world, the problem of the distribution of power tends to be secondary, since of necessity the power of decision does not lie with them, for all they might thwart decisions made. However, within current generations, environmental conflicts are often between those who have the economic, political and social power to shape economic projects and their outcome, and those who lack that power and who are driven to resist their outcome. The conflicts around what Martinez-Alier call the environmentalism of the poor are in particular of this nature. Similarly, the problems of social recognition are also to the fore. In particular, much of the recent dispute concerns not just the social invisibility of the poor, but more specifically, the lack of social recognition of forms of labour such as domestic labour and subsistence agriculture that fall outside the market realm. Within a market system, those forms of labour which have been central to sustaining human life and agricultural biodiversity tend to go unrecognised and unvalued (Martinez-Alier, 1997, 2002; cf. Shiva, 1992). The language of monetary valuation itself hides the other values and interests that characterise the ways environments matter to communities. Environmental conflicts concern not only unequal outcomes, but also the very language of valuation that is used to determine those outcomes (Martinez-Alier, 2002). Hence the resistance to the language of monetary valuation, as we already saw in Chapter 1 with respect to the Narmada Valley dam, is part of the wider struggle over the unequal distribution of environmental goods and harms.

Issues of power and recognition also play out within the environmentalism itself. In particular it plays a central role in the political controversies that surround the concept of wilderness in environmental discourse. The concept of wilderness played a problematic role in the colonial history of land appropriation. It still plays a part in the continuing appropriation of the lands of marginalised communities. In both cases it is premised on a failure to recognise the material impacts of the work of those communities within a landscape, or the cultural and social significance of environments to those communities. Their activities and cultures are rendered invisible. The following two chapters deal with these issues in more detail.[14]

Part III

Bringing environmentalism in from the wilderness

7 Wilderness, cultivation and appropriation

7.1 Wilderness and its critics

For a visitor to the United States a striking feature of some of the nature parks is the stark juxtaposition of commerce and parks. Shopping malls and strips often run up to the borders of a park. There they stop. Beyond one enters areas designated as wilderness, and as one does so one moves into a different institutional order. The market appears to run up against a very real boundary. Legally, the US wilderness designation partly does serve as a barrier to the expansion of commercial society. The legal designation of an area as wilderness under the Wilderness Act 1964 specifically includes limitations on commercial activities: 'Except as specifically provided for in this Act, and subject to existing private rights, there shall be no commercial enterprise . . . within any wilderness area designated by this Act' (Wilderness Act Section 4c). Wilderness in this context is understood as a place untouched by human interference, or as the Wilderness Act puts it – 'an area where the earth and its community of life are untrammelled by man, where man himself is a visitor' (Wilderness Act Section 4c).[1]

Wilderness thus understood has remained a central concept employed by environmentalists in the new world context of the United States. 'Nature' and 'wilderness' are the central normative categories of the deep green environmental movement. These attitudes are not, however, confined to explicitly deep green radicals. Something of them pervades the nature conservation movement. Natural landscapes, pristine and untouched by the marks of human beings, form the primary objects of appreciation. This is sometimes expressed in quasi-scientific terms, for example, in references to landscapes that exhibit 'ecological integrity' characterised by states that show only minimal human intervention.

At least one major motivation for that defence of wilderness has been to protect parts of the natural world from unregulated commerce. It is apparent, for example, in John Muir's much cited description of his opponents in the conflicts around the Hetch Hetchy Dam: 'These temple destroyers, devotees of ravaging commercialism, seem to have a perfect contempt for Nature, and instead of lifting their eyes to the God of the mountains lift them to the Almighty Dollar' (Muir, 1912, p. 716). Conversely, defenders of unregulated markets in the United States have typically opposed such wilderness designations and the very appeal to the concept of wilderness. For those in the 'Wise Use Movement' and 'Property Rights

Movement', the appeal to wilderness is likewise seen as a block to the proper expansion of market activity and a restriction on the exercise of property rights (Brick and Cawley, 1996). Hence, for example, one of the movement's principal advocates, Ron Arnold, places deep ecologists alongside 'eco-socialists' and 'establishment interventionists' as opponents of free markets (Arnold, 1996). The arguments around nature and the limits of commerce in this context have then often focused on the wilderness.

However, the debate is more complex than this initial account may suggest. The concepts of 'wilderness' and 'nature', understood as that which does not bear the mark of human activities, do not only have their critics among defenders of the free market and property rights. The appeal to the concepts also has critics from other directions. Two lines of criticism are particularly noticeable in recent discussions. The first is that from a variety of constructivists, who deny there is something called 'nature' to be defended. Those constructivist lines of argument have been often used against appeals to nature and wilderness by environmentalists (Vogel, 1996).[2] The second line of criticism is that developed in the environmental justice movement and concerns the role of appeals to 'nature' and 'wilderness' in the appropriation of land of often socially marginal populations, in particular their control and exclusion for the creation of 'nature parks'.[3] The export of the wilderness ideal from the US has not been a happy one.

These two lines of argument, from constructivism and from justice, are often found together. They are, however, logically independent and need I think to be distinguished from each other. The first line of argument from a strong form of constructivism should I think be rejected. Many of the arguments offered for the position are simply fallacious. Consider the following examples:

> Nature per se does not exist . . . Nature is only the name given to a certain contemporary state of science.
>
> (Larrere, 1996, p. 122)

> It is fair to say that before the word was invented, there was no nature.
>
> (Evernden, 1992, p. 89)

> We have no basis for distinguishing between Nature and our own changing historically-produced representations of nature . . . Nature is a cultural product.
>
> (Cupitt, 1993, p. 35)

The arguments for the constructivist position implicit in these passages are typical of many. They are based on simple use-mention confusions which the judicious use of quotation marks would have avoided.[4] For example, even given the dubious assumption that 'nature is only the name given to a certain contemporary state of science', it does not follow that 'nature does not exist', any more than the claim 'John is a name of 4 letters', and 'John wrote this book' entails that 'a name of 4 letters wrote this book'. In the sentences 'nature is only the name given to a certain contemporary state of science' and 'John is a name of 4 letters' the terms 'nature' and 'John' are mentioned, and strictly speaking should appear in quotation marks. In the sentences 'nature does not exist' and 'John wrote this book', the terms 'nature' and 'John' are used. Larrere's inference is a simple fallacy. Likewise Evernden's claim, 'It is fair to say that before the word was invented, there was no nature', is open to the charitable but banal reading that 'before the word "nature" was invented nothing was called "nature"', or the uncharitable reading 'before the word "nature" was invented there was no nature', which gets again any spurious plausibility it might have from a use-mention confusion. Compare the true but banal claim 'before the word "dinosaur" was invented nothing was called "a dinosaur"' and the clearly false claim 'before the word "dinosaur" was invented there were no dinosaurs'. Dinosaurs existed before 1841 (the date the term 'dinosaur' was coined by Richard Owen). These arguments against the appeal to 'nature' or 'wilderness' should be rejected.

However, the second line of argument against 'wilderness' which calls upon the association of the concept of wilderness with the appropriation of the land of the socially marginalised communities that live in wilderness areas is independent of these strong forms of constructivism. In so far as it maintains that areas are not wilderness it need not appeal to any strong constructivist claims. It can appeal more plausibly to claims about the history of those areas themselves and the lives of communities embodied in them. My own view is that the second line of argument is broadly right and my main purpose in this chapter will be to contribute to it by placing recent appeals to 'wilderness' in the context of a longer history of use of wilderness to justify the appropriation of land. In doing so we will see that the historical relationship between wilderness and the spread of commercial society is somewhat more complex than consideration of recent debates suggest. The wilderness designation had a central role in early modern political thought not as a means of protecting nature from commerce, but as means of demonstrating the need to bring it into the world of commercial relations. After developing some sceptical thoughts on wilderness, I finish by adding some qualifications through consideration of what survives of modern environmentalism after proper criticism of the wilderness ideal has been made.

7.2 All the world was America

The appeal to wilderness to appropriate land is not new to political argument. It has a long history in social and political thought. Indeed that history points to part of the problem with the wilderness ideal in deep green environmentalism. Two initial points need to be made here. First, the image of much of the 'new world' as wilderness relies upon a colonial perception of European settlers of an unspoilt pristine terrain dramatically different from the domesticated environments of Europe. Second, this image was associated with claims that were made for the justifiable appropriation of that land from its native population. Both points are illustrated well in the work of Locke and the characterisation of America as wilderness.

Consider Locke's comparisons of the 'wild woods and uncultivated waste of America left to Nature without any improvement, tillage or husbandry' (Locke,

1988, 2.37) with the improved and cultivated lands of the Britain and his corresponding account of the original appropriation of land:

> Whatsoever he tilled and reaped, laid up and made use of, before it spoiled, that was his particular Right; the Cattle, and Product was also his. But if either the Grass of his Inclosure rotted on the Ground, or the Fruit of his planting perished without gathering, and laying up, this part of the Earth, notwithstanding his Inclosure, was still to be looked on as Waste, and might be the Possession of any other.
>
> (Locke, 1988, 2.38)

Locke's references to the 'wild woods and uncultivated waste of America left to Nature without any improvement, tillage or husbandry' needs to be read in its historical context. It formed part of the denial of rights in land to the Aboriginal population. Locke, through the patronage of the Earl of Shaftesbury, was the secretary to the Lord Proprietors of Carolina and the Council of Trade and Plantations. Locke's theory of property was shaped by the arguments of colonialists for an ethical justification to appropriate the land of the native American population, particularly justifications that appealed to the rational use of land that had remained wild and unimproved. Subsequently appeal was made to Locke's arguments to justify further appropriation of native land.[5]

Locke's justification of private property in land given the Christian premise of original common ownership is voiced in terms of appropriation through cultivation and enclosure.[6]

> God, who hath given the world to men in common, hath also given them reason to make use of it to the best advantage of Life, and convenience . . . And though all the Fruits it naturally produces, and Beasts it feeds, belong to Mankind in common . . . yet being given for the use of Men, there must of necessity be a means *to appropriate* them some way or other before they can be of any use, or at all beneficial, to any particular Man. The Fruit or Venison which nourishes the wild *Indian*, who knows no Inclosure, and is still a Tenant in common, must be his.
>
> (Locke, 1988, 2.26, emphases in the original)

To claim that the 'wild *Indian* who knows no Inclosure' is still a tenant of the common is to deny his claims to the land. Indigenous populations had rights only to that which they had appropriated through their labour, and given the European image of them as hunters and gatherers which Locke reiterates here, appropriation extended only to that they caught and collected. The land being a wilderness, uncultivated and unenclosed, the 'vacant places of *America*' could be rightfully settled by Europeans without the consent of previous inhabitants or with their having 'reasons to complain' (Locke, 1988, 2.36).

> God gave the World to Men in Common, but . . . it cannot be supposed He meant it should always remain common and uncultivated. He gave it

> to the use of the Industrious and Rational (and *Labour* was to be *his Title* to it).
>
> (Locke, 1988, 2.34, emphases in the original)

The right to the original acquisition of land is subject in Locke's theory to two well-known limiting provisos, that produce not spoil and that there be good enough left for others. Both provisos are overcome through the introduction of money which can be hoarded without spoiling and which, through commerce, serves to foster the improvement of agriculture and hence 'to increase the common stock of mankind' (Locke, 1988, 2.37). The appropriation of America is justified by its being brought into the world of commerce and hence cultivation. It is the absence of money and commerce that explains the lack of productive appropriation of the land in America:

> What would a Man value Ten Thousand, or an Hundred Thousand Acres of excellent *Land* . . . in the middle of the in-land Parts of *America*, where he had no hopes of Commerce with other Parts of the World, to draw *Money* to him by the Sale of the Product? It would not be worth the Inclosing, and we should see him give up again to the wild Common of Nature, whatever was more than would supply the Conveniences of Life, to be had there for him and his Family . . . Thus, in the beginning, all the World was *America*, and more so than that is now; for no such thing as Money was anywhere known.
>
> (Locke, 1988, 2.48–2.49, emphases in the original)

By being brought into the world of commerce the wild common of America, uncultivated by the indigenous population is turned into a productive resource cultivated by the industrious and rational. The wilderness designation in this context served not to sustain nature outside of the world of commerce, but rather to bring it into commercial relations.

The Lockean account of the 'vast wilderness' of America as land uncultivated and unshaped by the pastoral activities of the indigenous population formed part of the justification of the appropriation of native land. It is also false. The land had been shaped by its native populations. However, it is a myth that has survived and has had lead to conservation management policies that ignore the impact of indigenous pastoral and agricultural activities. The problems are evident in the well-discussed problems in the history of the management of one of the great symbols of American wilderness, Yosemite National Park. In the influential report of the Leopold Committee, *Wildlife Management in the National Parks*, we find the following statement of objectives for parks:

> As a primary goal we would recommend that the biotic associations within each park be maintained, or where necessary recreated, as nearly as possible in the condition that prevailed when the area was first visited by the white man. A national park should represent a vignette of primitive America.
>
> (Leopold et al., 1963, p. 4., cited in Runte, 1987, pp. 198–199)

What was that 'primitive' condition? The experience of the first white visitors to the area is reported thus:

> When the forty niners poured over the Sierra Nevada into California, those who kept diaries spoke almost to a man of the wide-space columns of mature trees that grew on the lower western slope in gigantic magnificence. The ground was a grass parkland, in springtime carpeted with wildflowers. Deers and bears were abundant.
>
> (Leopold et al., 1963, p. 6., cited in Runte, 1987, p. 205)

However, this 'original' and 'primitive' state was not a wilderness, but a cultural landscape with its own history. The 'grass parkland' was in part the result of the pastoral practices of the Native Americans, who had used fire to promote pastures for game and black oak for acorn. After the Ahwahneeche Indians were driven from their lands by Major Savage's military expedition of 1851, 'Indian style' burning techniques were discontinued and fire suppression controls introduced. The consequence was the decline in meadowlands under increasing areas of bush. When Totuya, the granddaughter of chief Tanaya and sole survivor of the Ahwahneeche Indians was evicted from the valley, returned in 1929 she remarked on the landscape she found: 'Too dirty; too much bushy' (Olwig, 1995). It was not just the landscape that had changed. In the Giant Sequoia groves the growth of litter on the forest floor, dead branches and competitive vegetation inhibited the growth of new Sequoia and threatened more destructive fires. Following the Leopold report, both cutting and burning were used to 'restore' Yosemite back to its 'primitive' state.

However, to talk of restoring a park to a 'primitive', 'natural' or 'wilderness' condition simply disguises the nature of the problems. Reference to wilderness suppresses one part of the story that can be told of the landscape: the non-European native occupants of the land are themselves treated as part of the 'natural scheme', of the 'wilderness', without history and not as dwellers in a landscape which embodies their own cultural history. Moreover, the language also disguises the way in which the history of the landscape is being artificially frozen at a particular point in time: 'The goal of managing the national parks and monuments should be to preserve, or where necessary recreate, the ecologic scene as viewed by the first European visitors' (Leopold et al., 1963, p. 21, cited in Runte, 1987, p. 200). To refer to mythical 'natural' or 'wilderness' states avoids the obvious question 'why choose that moment to freeze the landscape?' There are obvious answers to that question, but they have more to do with the attempt to create an American National culture than they do with ecology. And here there are normative objections to made of the use of wilderness to gloss over part of the darker side of American history. The appeal to wilderness, in this context at least, is a way of avoiding coming to terms with a troubling historical dimension to environmental evaluation.

The shift to the land management policies embodied in the Leopold report with their attempt to restore landscapes to their 'primitive' pre-European condition from those found in the work of Locke also reflects a change in attitudes to wilderness.

The wilderness ideal is itself historically a local one. It is uncultivated land that the environmentalist now attempts to protect, not the cultivated landscapes that Locke praises. As we have already noted in Chapter 3, the dominant perceptions of land and landscapes have shifted. Recent environmental thought has echoed Mill's romantic influenced observation about the limits of agricultural expansion quoted earlier.

> It is not good for man to be kept perforce at all times in the presence of his species. A world from which solitude is extirpated, is a very poor ideal. Solitude, in the sense of being often alone, is essential to any depth of meditation or of character; and solitude in the presence of natural beauty and grandeur, is the cradle of thoughts and aspirations which are not only good for the individual, but which society could ill do without. Nor is there much satisfaction in contemplating the world with nothing left to the spontaneous activity of nature; with every rood of land brought into cultivation, which is capable of growing food for human beings; every flowery waste or natural pasture ploughed up, all quadrupeds or birds which are not domesticated for man's use exterminated as his rivals for food, every hedgerow or superfluous tree rooted out, and scarcely a place left where a wild shrub or flower could grow without being eradicated as a weed in the name of improved agriculture.
>
> (Mill, 1994, Book IV, ch. 6, section 2)

It is something like Mill's Romantic-influenced vision that informs the modern environmental movement.

However, the problem with this wilderness ideal is that it is not just an historically local perception that was associated with the appropriation of land. It is socially and geographically local as well and it retains its link with appropriation. However, the appropriation is now made in the name of wilderness rather than cultivation. It is invoked in the creation of 'nature parks' for the new eco-tourism, at home and abroad, which is premised on the assumption that nature requires at least the absence of the marks of human activity and at best the absence of people. It has led to policies of exclusion and control of the indigenous human populations on the grounds that they do cultivate and shape the land. Thus the development of conservation parks in the developing world through the eviction of the indigenous populations that had previously lived there. In Africa consider the fate of some of the Masai who have been excluded from national parks across Kenya and Tanzania.[7] Attempts to evict indigenous populations from the Kalahari reveal the influence of the same wilderness model: 'Under Botswana land use plans, all national parks have to be free of human and domestic animals'.[8] The history of exclusion is illustrated in the conflicts surrounding the Batwa in Uganda. They were 'officially' excluded from forest reserves during the British colonial period of the 1930s, although in practice they continued to use forests as a means to livelihood. Since the establishment of National Parks in 1991, their exclusion was made effective, which has led to continuing conflicts (Griffiths and Colchester, 2000).

Similar stories are to be found in Asia where the alliance of local elites and international conservation bodies has lead to similar pressures to evict indigenous populations from their traditional lands. In India, the development of wildlife parks has lead to a series of conflicts with indigenous populations. In India there has been a series of much discussed evictions and resettlements of local populations in the creation of parks and sanctuaries. Consider for example the resettlements of Maldheris in the Gir National Park (Choudhary, 2000), the proposed and actual exclusions of local populations in the Melghat Tiger Reserve and Koyna Sanctuary in Maharashtra, and the conflicts around the Nagarhole National Park where there have been moves from the Karnataka Forest Department to remove 6,000 tribal people from their forests on the grounds that they compete with tigers for game (Guha, 1997; Griffiths and Colchester, 2000; Jayal 2001). The moves are supported by international conservation bodies. Hence the remark of one of experts for the Wildlife Conservation Society in the Nagarhole case – 'relocating tribal or traditional people who live in these protected area is the single most important step towards conservation' (cited in Guha, 1997, p. 17).

The control and exclusion of populations is a theme that runs through much anthropological work on the environment. Consider the comment from a person in the Makala-Barun National Park and Conservation Area in Nepal reported by Ben Campbell: 'This park is no good. They don't let you cut wood, they don't allow you to make spaces for paddy seed-beds, they don't permit doing *khoriya* [a form of slash and burn agriculture]' (Campbell, 2005, p. 333). The wilderness model fails to acknowledge the ways parks are not wilderness but a home for its native inhabitants, the degree to which the landscapes and ecology of the 'wilderness' were themselves the result of human pastoral and agricultural activity, and the cultural significance of particular landscapes, flora and fauna to the local populations. In so far as the indigenous populations are recognised they are often themselves treated as a kind of exotic fauna, who are a part of nature, rather than fellow humans who also transformed their landscapes.

The conflicts between the attempts to create nature parks that embody the wilderness, and the local often marginalised populations who have lived and worked in that wilderness are not confined to the developing world. Consider, for example, the following comments of a local living by the natural park of Sierra Nevada and Alpujurra granted biosphere status by UNESCO and Natural Park status by the government of Andalusia:

> [Miguel] pointed out the stonework he had done on the floor and lower parts of the wall which were all made from flat stones found in the Sierra. I asked him if he had done this all by himself and he said 'Yes, and look, this is nature' ('Si, y mira, esto es la naturaleza'), and he pointed firmly at the stone carved wall, and he repeated this action by pointing first in the direction of the Sierra [national park] before pointing at the wall again. Then, stressed his point by saying: 'This is not nature, it is artificial (the Sierra) this (the wall) is nature' ('Eso no es la naturaleza, es artificial (the Sierra) esto (the wall) es la naturaleza).'
>
> (Lund, 2006, p. 382)

In the humanised landscapes of Europe, the concept of 'wilderness' as such has little appeal, a fact that is sometimes acknowledged by European conservationists with something of an inferiority complex. Proper nature is elsewhere. The fact is often viewed by the visiting deep ecologist with incredulity at the ways in which landscapes in Europe are managed. Here, for example, is a comment from the late Richard Sylvan on his visit to the Three Peaks area of the Yorkshire Dales in the United Kingdom.

> [T]he Three Peaks district is now prized for its recreational values, it is prized for its *comparative remoteness and wilderness*, its fewness of people and absence of industry, for the walks and wild meadows it offers. But it is a landscape far removed from its *pre-agricultural original*. It has been almost totally *stripped of its native vegetation*, and most habitats and much of *its ecology destroyed, the remainder substantially modified*, in the former quest . . . for agricultural advantage and optimal, or often excessive, grazing usage. The district remains starkly treeless . . . But woods, formerly with different wildlife, there formerly were, as a tiny protected strip at Colt Park pleasantly testifies. Most of the district still remains overrun by, and severely eroded by, sheep, which none but subsidized and distorted market system would support. Remarkably, however, there appears little pressure for economic adjustment and ecological restoration, for removal of some sheep and return of more woods. Many recreationalists appear to prefer impoverished grasslands, treelessness. Even environmental organisations like English Nature own sheep and lease out lands for sheep grazing.
>
> (Sylvan, unpublished, emphases added)[9]

What is notable here, as the passages I have emphasised here indicate, is that the ideal by which landscapes are to be judged is the unmodified 'pre-agricultural original'. It is an ideal of nature independent of human intervention that forms the standard from which others are judged.

Conservationist organisations like English Nature do not aim to at anything like Sylvan's 'pre-agricultural original'. However, while there is no appeal to wilderness in the UK context, real conflicts do exist in the perceptions of the landscapes between farmers and conservationists (Walsh et al., 1996; O'Neill and Walsh, 2000). Farmers on the one hand and landscape planners and conservationists on the other have different perceptions of what constitutes a good environment of which some are self conscious. As we noted in Chapter 3, farmers' perceptions are often husbandry based. Hence the comment of a farmer in the Yorkshire Dales: 'A farmer will look at someone else's farm and could tell whether it was well farmed or not. They wouldn't look at the view and think '"What a good view!"' (Walsh et al., 1996, p. 22).[10] Given that perspective the wildness loved of the conservationist can be seen as a defect.

> If a piece of land's conserved, it tends to get overgrown, it gets brown. I suppose people from off will tend to look at that and admire its tones, in autumn sort

> of thing. Or golden spring, or whatever. But a farmer will look at it and think – it's overgrown.
>
> (Walsh et al. 1996, p. 22)

The farmer will sometimes look upon the land in a way that is different from both the nature conservationist and the visitor admiring the landscape. Attempts to fence off nature and allow it to grow wild are met with disapproval: it represents a 'mess' (Walsh et al., 1996, p. 23). Hence, the resistance felt by some farmers to the authorities who represent conservationists and landscapers – outsiders who aim to mould the environment in ways that are alien to their own husbandry based conceptions. Correspondingly there is the articulated threat to a community founded upon farming being transformed into a museum exhibit to conform to some idealised Romantic image of how the countryside should look: 'National Parks, English Nature, they'll finish up with all the farmers running about in smocks, like museum curators. That's not a community. We have a community which is a working community' (Walsh et al., 1996, p. 44). The worry here is that a particular conception of the way nature and landscape ought to be is being imposed from the outside on those who live and work in an environment, for whom nature is not primarily an object of scientific interest or aesthetic contemplation, but something with which one has a working relationship.

7.3 What's left of the wilderness?

Thus goes the case for the prosecution against the wilderness ideal. It is an historically and socially local vision, historically implicated in the colonial appropriation of land, and currently implicated in the exclusion or control of often poor and powerless groups that live in lands too marginal to sustain intensive agricultural activity. The case against the wilderness model in this context is I think powerful and one with which I broadly concur. It also might suggest, that perhaps we should reverse the order of primacy between natural and cultural landscapes, where cultural landscapes are seen as poorer relatives to their natural relatives. For it might be argued, that the natural landscape itself is just a particular cultural landscape, one that has a particular social and cultural history. Landscapes themselves, like that in the Yorkshire Dales or the Yosemite, are managed to mould them to expected patterns. And even where landscapes are not directly managed, the perception of the landscape as 'natural' or 'wilderness' is itself a culturally specific achievement. The conflicts outlined in section 7.2 are conflicts between different cultural landscapes. These can be direct material conflicts on how landscapes themselves should be transformed human activity. However, they can also be conflicts in ways of seeing landscapes – consider the comments of the farmers above. Conflicts about appropriation can likewise be conflicts on who has legitimate powers to determine the material future of landscapes, with who has property rights and economic and political power to shape a landscape. However, conflicts about appropriation can also have a cultural and symbolic dimension, as to which perceptions and understandings of environments predominate.[11] The conflicts between the farmers and the conservationists in the Yorkshire Dales, or between Miguel and the park authorities in Andalusia, or between the peasants and park authorities in Nepal have both dimensions. They are in part about who has rights to direct and control the land, but they are also about how the land is to be described and perceived. On this view then environmental conflicts are conflicts between different cultural landscapes. Natural landscapes are cultural landscapes that dare not speak their name. Pushing the line of argument further, it is sometimes argued that we should drop the notion of nature altogether from environmental discussion (Vogel, 1996).

There is I think much in that line of argument. However, the final conclusion, that nature disappear altogether from discussion is I think mistaken. Neither would I want to reject the environmentalist's view as simply internally incoherent. Indeed I think there is much to the environmentalist's position that can be rescued from the wilderness. An initial point to be made here is that even appeal to 'wilderness' itself has a more ambivalent role in the politics of nature than my discussion this far might suggest. Wilderness, in particular in the romantic celebration of it, was sometimes appealed to justify public access to what is common against the privatisation of land. Consider, for example, Mill's comments on access to uncultivated land:

> [T]he exclusive right to the land for purposes of cultivation does not imply an exclusive right to it for purposes of access; and no such right ought to be recognised, except to the extent necessary to protect the produce against damage, and the owner's privacy against invasion. The pretension of two Dukes to shut up a part of the Highlands, and exclude the rest of mankind from many square miles of mountain scenery to prevent disturbance to wild animals, is an abuse; it exceeds the legitimate bounds of the right of landed property. When land is not intended to be cultivated, no good reason can in general be given for its being private property at all; and if any one is permitted to call it his, he ought to know that he holds it by sufferance of the community, and on an implied condition that his ownership, since it cannot possibly do them any good, at least shall not deprive them of any, which could have derived from the land if it had been unappropriated.
>
> (Mill, 1994, Book II, ch. 2, section 6)[12]

The appeal to wilderness in this context forms part of an assertion of rights to common access for common enjoyment.[13] Such appeals were central for example to the struggles in the United Kingdom for access to mountains and moorland by the urban working class in the nineteenth and twentieth centuries that culminated in the mass trespass movement. As we noted at the outset of this chapter, it still animates parts of current nature conservation which appeals to the need to maintain boundaries around land that protects it from privatisation and commercialisation. Hence the central worries of defenders of wilderness is that its critics remove constraints on the commercial development of currently protected areas. The wilderness designation of an area can matter. For the reasons I have outlined, I do not believe that the concept of wilderness in the sense of places relatively untouched by human intervention is the appropriate concept to use in this context. There is little if any wilderness in this sense. Neither am I convinced that employing the concept in its older sense of 'uncultivated land' fares particularly better. However, the defence of particular places and the maintenance of boundaries against commerce are entirely proper – and there are good reasons to hold that this should include places in which the albeit culturally specific experience of wildness is possible.

Moreover, much of the ethical vocabulary which environmentalists call upon to criticising features of some of our contemporary relations to the non-human world survives the rejection of the wilderness model: for example, reference to the cruelty inflicted on fellow creatures, of the failure of care involved in the wanton destruction of places rich in wildlife and beauty, of the pride and hubris exhibited in the belief that the world can be mastered and humanised, of our lack of a sense of humility in the midst of a natural world that came before and will continue beyond us. Nor do I think there are reasons to deny that both arts and sciences have developed the human senses in ways that allow humans to respond to the qualities that objects possess in a disinterested fashion to objects and that in doing so they have developed a human excellence (O'Neill, 1993c). What is I think true is that they have a particular local cultural origin.[14]

Does the fact that they have a particular cultural origin matter? There are certainly occasions when it appears to matter. Consider the following incident. Returning from a winter climbing trip in Glencoe in Scotland, two friends and myself were passing Loch Lomond. It was a day of bright sunshine and without wind. There was not a ripple on the Loch and the mountains were reflected without flaw in the water. Two of us made the kind of comments full of expletives you'd expect on such occasions. The third who had a training in the history of art then began an account of the development during the eighteenth and nineteenth centuries of the aesthetic responses to the landscape which we had just exhibited. And it completely ruined the moment. Should his comments have undermined our appreciation? Does knowledge of the cultural origins of our responses to the natural world destroy those responses? Clearly such histories can have that effect. It is the source of the power of genealogical criticisms of social practices and attitudes: the history of the use of the concept of wilderness outlined earlier is perhaps an example. Or to take another, once one has read the story of the highland clearances in Scotland, it is difficult not to see a depopulated landscape rather than a pristine wilderness.

However, the cultural self-understanding of our attitudes, understanding and perceptions of non-human nature need not undermine them. They can be seen as cultural achievements. Gellner comments that anthropologists sometimes tend to be liberals at home and conservatives abroad (Gellner, 1973, p. 29). The comment is not facetious. It raises important methodological issues for anthropology – most notably whether the suspension of criticism of those with whom one is engaged is a condition of understanding, and if not what the principles of interpretation ought to be (Gellner, 1973, p. 29). I leave these methodological issues aside here, however. I want to add a variant to the point that sometimes perhaps they may also have a tendency to be celebratory abroad and deflationary at home. Even if the responses to landscapes have historically and socially local origins, they are ones that can have their own virtues, that make a contribution to a wider conversation about values.

There are two points I think need to be made here. First, the cultural sources of our responses need to be distinguished from the objects of our responses. The point is one that needs to be stressed against certain strong forms of constructivism. For the strong constructivist, once we are made aware of the cultural origins of our responses we realise that there is no 'nature' there, that we are surrounded by a world of cultural objects. That strong constructivism is mistaken. There is a clear distinction to be drawn between the *sources* of our attitudes, which are economic, political and cultural, and the *objects* of our attitudes which can still remain non-cultural. That our capacity to appreciate and respond to the non-human natural world in a certain way is a cultural achievement, the outcome of social and cultural processes, does not entail that the object of our attitudes is a cultural object. This is not to deny that many landscapes that are presented as 'natural' are cultural in a real material sense – they are the result of human activity. Hence, the proper redescriptions of wilderness outlined earlier. However, it would simply false to hold that at the level of geology for example that *all* is a human product: it has a history before us. More generally, the world in which we live is the result of an interplay of human and non-human history. And at the level of processes rather than objects or end-states, we live in a world of unintentional non-human natural processes that proceed regardless of human intentions and indeed which often thwart them. This is a source not only of human sorrow, for example of life and land lost in flood, but also of human delight, for example at the plant or bird that arrives uninvited in an industrial wasteland. It is possible for culture to foster appreciation of non-human objects that themselves are not cultural – to maintain a sense of the otherness of non-human nature and our place within it. The picture of a world in which humans can see nothing but the reflections of themselves is itself a peculiar modern human conceit that our constructivist times tends to encourage. There is a core of environmentalism that is properly critical of that conceit. We live in a larger world of which human life is just a part (Goodin, 1992, ch. 2).

Second, the particular local cultural origins of responses and vocabulary with which particular groups approach the natural world does not entail that they cannot belong to a wider conversation. All assertions, both factual and normative, have a local origin – they could have no other. That is consistent with some making rational claims on a wider audience. However, whether they do so or not is clearly another matter. In particular the question of whether specifically ethical claims can make any claims on a wider audience raises important practical and ethical issues that deserve more detailed treatment. More specifically still, how far are the assertions of environmentalists merely expressions of some local mores and values that can make no claim beyond particular cultures and how far can they make more wider claims that transcend their particular origins? This question is one that I will address in Chapter 8.[15]

8 The good life below the snowline

8.1 Authoritarian environmentalism?

The title of this chapter was prompted by reading a few stanzas from a poem of W. H. Auden – 'Letter to Lord Byron'. Much in that poem represents a criticism of what Auden calls the 'excessive love for the non-human faces' and as such it anticipates some recent criticisms of the anti-humanism of some deep greens. However, this theme is not my central concern here. The stanzas that prompted the thoughts that follow were these:

> The mountain-snob is a Wordsworthian fruit;
> He tears his clothes and doesn't shave his chin,
> He wears a very pretty little boot,
> He chooses the least comfortable inn;
> A mountain railway is a deadly sin;
> His strength, of course, is as the strength of ten men
> He calls all those who live in cities wen-men.
>
> I'm not a spoil-sport, I would never wish
> To interfere with anybody's pleasures;
> By all means climb, or hunt, or even fish,
> All human hearts have ugly little treasures;
> But think it time to take repressive measures
> When someone says, adopting the 'I know' line
> The Good Life is confined above the snow-line.
>
> Besides, I'm very fond of mountains, too;
> I like to travel through them in a car;
> I like a house that's got a sweeping view;
> I like to walk, but not to walk too far.
> I also like green plains where cattle are,
> And trees and rivers, and shall always quarrel
> With those who think that rivers are immoral.
>
> (Auden, 1968, pp. 59–60)

Now I confess that I am one of Auden's mountain snobs, although I question the degree to which I'm the fruit of Wordsworth – poets, like the rest of us, are apt to exaggerate their own influence. My tastes about mountains differ from those of Auden. However, partly for that very reason, there are two themes in Auden's poem which I believe deserve serious consideration and which have been developed in more prosaic terms in more recent criticism of environmentalism.

We have touched on these themes already in Chapter 7. The first is the extent to which much in modern environmentalism might be taken to be a local vision. The second is the potentially authoritarian way this local vision is imposed on others. A particular conception of the good life, according to which the good life has to include a life above the snowline, in a wilderness devoid of human faces, is presented as a universal vision of the best life for humans. The worry is that much in modern environmentalism is simply the attempt to impose that local conception of the good life on others. Correspondingly there is sometimes a failure in modern environmentalism to appreciate the variety of ways in which different human relationships to the natural environment.

To capture the very real potential authoritarian consequences of environmentalism consider a second critical commentary on the Wordsworthian vision of nature, a 1929 essay of Aldous Huxley, 'Wordsworth in the Tropics'. Huxley opens with remarks on both the localism and influence of the Wordsworthian vision:

> In the neighbourhood of latitude fifty north, and for the last hundred years or thereabouts, it has been an axiom that Nature is divine and morally uplifting. For good Wordsworthians – and most serious-minded people are now Wordsworthians either by direct inspiration or at second hand – a walk in the country is the equivalent of going to church, a tour of Westmorland is as good as a pilgrimage to Jerusalem.
>
> (Huxley, 1929, p. 113)

Now, like Auden, Huxley exaggerates I think the influence of poets – but this I leave aside and for the moment use the term 'Wordsworthian'. The central claim of Huxley's essay is that the Wordsworthian vision would not travel well. It is possible only in the domesticated environments of Europe, not in the non-temperate zones.

> The Wordsworthian who exports this pantheistic worship of nature to the tropics is liable to have his religious conviction somewhat rudely disturbed. Nature under a vertical sun, and nourished by the equatorial rains, is not at all like that chaste, mild deity who presides over . . . the prettiness, the cosy sublimities of the Lake District . . . There is something in what, for lack of a better word, we must call the character of great forests . . . which is foreign, appalling, fundamentally and utterly inimical to intruding man.
>
> (Huxley, 1929, pp. 113–114)

For all Huxley's perceptive remarks in his essay on the conditions for certain kinds of appreciation of nature, his claim here is thoroughly mistaken. In the first place, subsequent developments have shown the Wordsworthian vision to be capable of travelling much better than Huxley assumes. It has translated itself well into the tropics. Modern nature lovers no longer confine themselves to the Lake District or the Alps. The trip to the tropics has become part of the Wordsworthian pilgrimage and a whole tourist industry exists dedicated to the transport believers to non-temperate zones. Nor is this export of the Wordsworthian vision a recent development. Indeed, Huxley's own account of alien nature of the tropics is part of the export industry. His own view is much closer to the Wordsworthian vision than he assumes. And this takes me to a second point about Huxley's characterisation of the tropics as places of wilderness which are 'fundamentally and utterly inimical to intruding man' and which are in stark contrast with the domesticated environments of Europe – that characterisation is drawn in European colours. The account renders invisible all those persons for whom the tropics were not and are not an alien and impenetrable wilderness, but their home.

And here again the potentially intolerant and authoritarian dimensions of the environmentalist vision is apparent. Not only is Huxley's Wordsworthian doing well in the tropics, but also there nature itself is being remoulded to confirm with his image of it as a place in which humans have no place. As we noted in Chapter 7, nature parks are created and legitimised through the wilderness model of nature conservation which puts considerable emphasis on the values of wilderness, understood as nature untouched by humans, and of 'ecological integrity' understood as the integrity of ecological systems from human interference. This wilderness model developed historically through that image of tropical nature invoked by Huxley as an unspoilt wilderness that contrasts with the domesticated environments of Europe (Anderson and Grove, 1987; MacKenzie 1988; Grove, 1995). It was a concept to which appeal was made in the appropriation of native land and to which appeal is still made. However, as we noted in Chapter 7, appropriation has shifted from being justified in terms of the failure of indigenous populations to cultivate the land, a justification that we have seen was central to the work of Locke, to being justified on the basis of their failure to leave it uncultivated.

The shaping of non-temperate nature in conformity with the European image of a wilderness is often achieved through the coercive measures against those whose home the wilderness is. Conservation and nature parks in the developing world are developed through the exclusion and control of the indigenous populations who had previously lived there. Hence the conflicts between the attempts to create nature parks that embody the wilderness and the local and often marginalised populations who have lived and worked in those parks. Neither are these conflicts confined to the tropics. As we noted in Chapter 7, similar resistance is to be found among those faced by exclusions or controls in the cause of nature within Europe. Indeed back at home in the more traditional places of Wordsworthian pilgrimage such as the Lake District and the Yorkshire Dales, similar conflicts are played out in a less dramatic way. These too have been shaped to produce a landscape that conforms to a particular Romantic aesthetic. Consider for example the conflicts outlined in Chapter 7 between farmers and landscape planners and conservationists in the Dales National Park in Yorkshire over the perceptions of what constitutes a good landscape. In all these cases the conflicts raise real issues about the ways in which nature conservation involves the imposition of a historically and socially local romantic conception of the way we should view and shape our environments on populations whose communities are constituted by a distinct working relationship to those environments.

8.2 Liberal environmentalism?

How might one respond to these potentially authoritarian features of environmentalism? One response is that of a particular anti-perfectionist version of liberalism. On this account the problems just outlined are what is to be expected given a particular perfectionist conception of politics and public policy that is assumed by a great deal of green political theory and practice. On this perfectionist conception the aim of politics is to foster the good life. The conception has a long ancestry. Its classic statement is to be found in Aristotle: 'the end and purpose of the polis is the good life' (Aristotle, 1948, 1280b 38f) where the good life is characterised in terms of the virtues: hence the comment that the best political association is that which enables every man to act virtuously and to live happily (Aristotle, 1948, 1324a 22). Modern green political theory adds a particular dimension to that account: a properly constituted relationship with the natural world becomes part of the good life. Consider, for example, Nussbaum's claim that 'a creature who did not care in any way for the wonder and beauty of the natural world' (Nussbaum, 1990, p. 222) would lack an element that is constitutive of what it is to be human. Or to take a formulation from my own previous work: 'the best human life is one that includes an awareness of and practical concern with the goods of entities in the non-human world' (O'Neill, 1993a, p. 24).

For the liberal, the authoritarianism of the environmentalist merely exhibits just what is wrong with this Aristotelian account of the public life. The classical view of politics according to which 'the end and purpose of the polis is the good life' is incompatible with the pluralism of modern life. Specifically an environmentalism which is founded upon the premise that some particular form of relationship to nature is a component of the good life is founded upon a failure to recognise the variety of ways a good life can be lead without the particular relationship to nature proffered as best. There is a plurality of different ways a good life can be led below the snowline involving a relation to nature which is very different from that advocated by many of the defenders of environmentalism who appeal to some universal shared appeal to the 'wonder and beauty' of the natural world. There are individuals who live perfectly good flourishing lives below the snowline, for whom nature is a place for their working life, or who, like Auden, love the life of the city and are content with the nature of the garden or the city park or the view from the car. The objection runs that, in so far as the modern green movement does involve a kind of Aristotelian perfectionism, it issues in just the kind of authoritarianism outlined in section 8.1. It leads inevitably to the coercive imposition of a particular vision of the good life on others who do not share it, whether this be within the confines of a particular state or through the missionary activities of its proponents across states. Against the background of the plurality of ways in which a good life can be lead in the modern world, an appeal to a thick theory of the good in public life necessarily issues in the imposition on others of a particular local conception of the good. Modern history has shown the appeal to any such conception of the good life is at best authoritarian and unjust in its implications, at worst totalitarian.[1] Public policy should be neutral between different conceptions of the good.

This particular liberal perspective itself breaks with the perfectionism of classical liberalism of the form defended by J. S. Mill's. As we noted in Chapter 4, for Mill, 'the first question in respect to any political institutions is, how far they tend to foster in members of the community the various desirable qualities moral and intellectual' (Mill, 1975, p. 29). Mill (as we noted in Chapter 7) was also happy to include a particular relation to the natural world as part of the human good: 'It is not good for man to be kept perforce at all times in the presence of his species' (Mill, 1994, Book IV, ch. 6, section 2). The appeal to liberal neutrality also breaks with the central traditions of republican and socialist thought. However, the defence of liberal neutrality has been influential in recent discussion. Moreover, as I noted in Chapter 4, it offers one set of grounds for starting with market-based approaches to environmental choices which are purely want-regarding, be this cost-benefit analysis or the development of markets in environmental goods and harms.[2]

The appeal to liberal neutrality is not incompatible with all environmental concerns. Justification of public policy can appeal to a thin account of the goods of human life in the sense that Rawls uses the term. That is, it can appeal to those primary goods that any rational agent will want whatever their particular conception of the good life. These will include many basic environmental goods. Thus, for example, in so far as environmental politics is concerned with issues of resources and pollution that are threats to primary goods they fall within the ambit of a liberal political order. However, once environmental politics goes beyond those kinds of domain and argues for some specific thick theory of the good it involves a perfectionist politics that is inconsistent with the pluralism of the modern world. A liberal environmental politics has to be written in a thin ethical language that does not presuppose specific local conceptions of the good life.

This liberal argument for resisting any move beyond any public ethic that is written in a thin ethical vocabulary is sometimes buttressed by appeal to a wider set of enlightenment emancipatory values which are taken to involve taking a critical standpoint that transcends what is local. Local cultural practices are sometimes oppressive to particular individuals and groups, for example to women and subordinate castes and classes. To launch criticism of those practices it is argued that one needs a standpoint and set of normative concepts that transcends the local, that allows the possibility of standing outside particular social practices to formulate sceptical questions about them. The argument runs that such a standpoint requires a minimal moral language – an ethical discourse written in terms that have 'thin' or minimal meanings, that employ general abstract terms of rights and goods that are taken to be universal, transcending the specific ethical understandings of local culture, in contrast with thick concepts whose understanding requires immersion in particular local practices. A thin vocabulary of the right and the good is a condition of being able to raise a question about specific moral concepts, for example, of being able to ask with Nietzsche of the virtue of humility – 'Is humility good?'[3]

What might be said in response to these arguments for a liberal moral minimalism in public discourse? An initial and fairly standard response is that liberal moral minimalism itself rules out too much and in consequence that it carries its own form of authoritarianism and injustice. It silences in the public sphere certain forms of ethical language and valuation, often associated with marginalised voices, and privileges others. Thus in the specifically environmental sphere it might be argued that it leads to an impoverishment of the kinds of language that can be appealed to in public policy debates which have been a concern of this book. Environmental policy documents are dominated by an evaluative language written in either an economic or quasi-scientific terms. Environmental values are reduced to preferences that can be captured by monetary valuations. Environmental policy goals are reformulated in economic terms – hence the ways in which sustainability is recast in the language of capital. The values that survive that treatment are those that can be written terms that have associated with them some kind of scientific authority – consider for example the special role that 'biodiversity' holds in national and international policy documents. Part of the justification for the use of such terms is their neutrality between particular thick conceptions of the good and thick normative frameworks. However, as we have seen in Parts I and II of this book, those terms do not capture the kinds of concern that move responses to the loss of environmental goods. Indeed, those responses are rather often expressed through a resistance to the terms in which the public debate is reformulated. Hence the resistance to the attempts to capture environmental concerns in monetary terms we noted in Parts I and II. As I noted in the introduction to this book, this is true more widely of resistance to the globalisation of the domain of markets. Resistance is articulated in terms of thick understandings of the goods in question. No minimal moral language is able to capture the terms of those response to the spread of market norms. The attempt to reduce the language of public discourse to some thin moral language carries then its own forms of marginalisation of voices. As Joan Martinez-Alier has argued, a significant dimension environmental justice is the question of which language of valuation is to be employed in disputes: 'Who has the power the power to simplify complexity, ruling some languages of valuation out of order?' (Martinez-Alier, 2002, p. 271).

Similar points apply to the resistance to nature conservation policy discussed in Chapter 7. Those policies are often blind to the local understandings of the values of those environments. The attempt to recover and articulate such thicker local vocabularies and understandings that individuals and groups bring to the environments has been at the centre of much recent environmental anthropology some of which I alluded to in that discussion. That this anthropological work moves in the opposite direction to the minimalism of some recent variations of liberal moral philosophy is not surprising. The recovery of thick descriptions is sometimes taken to define the anthropological enterprise. Clifford Geertz, for example, in *The Interpretation of Cultures*, characterises anthropological analysis in these terms:

'What defines [anthropological analysis] is the kind of intellectual effort it is: an elaborate venture in, to borrow a notion from Gilbert Ryle, "thick description"' (Geertz, 1973, p. 6). On the line that Geertz develops, the exercise in uncovering layers of interpretative depth defines the anthropological enterprise. In doing so it has or can have a political and ethical agenda of its own. While anthropology has its own complex relation with colonialism, recent critical environmental anthropology often characterises itself, at least implicitly, as representing the voice of local and often marginalised groups against the dominant global discourses. The languages of valuation employed in the alliance that is sometimes forged between nature conservation bodies and international centres of power against local actors are typical in this regard. Such global normative discourse far from fulfilling the promise of enlightenment cosmopolitanism and offering a standpoint to criticise illegitimate power and injustice embodied in local social practices promised by enlightenment cosmopolitanism, becomes rather a way in which global economic and political power is itself expressed.

However, to make these points about this anthropological project also highlights a worry that underpins one attraction of moral minimalism, that the project of recovering thicker descriptions leads to a form of relativism that is inimical not just to the cosmopolitan ambitions of the enlightenment but to any social criticism. This anthropological venture in thick descriptions is often associated with a rejection of any kind of universalism about ethical claims. Thick understandings, it is often assumed, are local and specific to particular cultures. They leave no room for external criticism of social practices. Hence Gellner's (1973) charge noted in Chapter 7, that anthropologists are often liberals at home, but conservatives abroad. The worry about relativist versions of the venture in recovering thick descriptions is one that I share. Indeed, I do not think the venture can survive as a critical project thus understood. Anthropological work which has a political position in defence of local cultural understandings implicitly calls upon the same emancipatory values of the enlightenment as its opponents – and it is none the worse for that. The arguments about the ways in which particular voices and perceptions are silenced has critical power only to the extent it calls upon shared beliefs about the values of equality in standing, voice and power.

The problem in the debate lies in the shared assumption that any venture towards thick ethical description is at the same time a shift away from universal values to those that are purely local and hence unable to make claims on those outside a particular shared set of practices. The problem here can be approached through an ambiguity in the way the terms 'thick' and 'thin' are themselves used in recent philosophical and anthropological literature. The terms are used in a number of distinct senses which are often elided:

1 *Interpretative depth*: Geertz's (1973) own use of the term has origins in Ryle's work in the philosophy of mind. Ryle (1971a) used the terms thick and thin descriptions in part to capture differences in the interpretative depth of action descriptions. The thinnest description of an action is a description of physical behaviour:

> Two boys fairly swiftly contract the eyelids of their right eyes. In the first boy this is only an involuntary twitch; but the other is winking conspiratorially to an accomplice. At the lowest or thinnest level of description the two contractions of the eyelids may be exactly alike. From a cinematograph-film of the two faces there might be no telling which contraction, if either, was a wink, or which, if either, was a mere twitch. Yet there remains an immense but unphotographable difference between a twitch and a wink.
>
> (Ryle, 1971a, p. 480; cf. Ryle, 1971b)

Descriptions are thickened as interpretative depth is added. Hence Ryle's examples of successively thicker descriptions: a boy contracts his eyelid; a boy is winking; a boy is parodying another's attempt at winking; a boy is practising a parody of another's attempt at winking. Thickness in the sense refers then to the interpretative depth of descriptions of behaviour.

2 *Theories of the good*: Rawls (1972) uses the terms to draw a distinction between theories of the good, between a thin theory which refers to the good 'restricted to the essentials' and determines the class of primary goods that rational individuals will necessarily require to pursue whatever ends they might have, and a thick theory of the good which specifies particular ends (Rawls, 1972, section 60).

3 *Specific versus general ethical concepts*: Another usage of the terminology, due largely to Williams (1985, pp. 129–131 and 140ff.), is to draw a distinction between ethical concepts like brave, cowardly, kind, pitiless, which are specific reason-giving concepts and world-guided in the sense that their application is 'determined by what the world is like', and ethical concepts, like good, bad, right and wrong which are general, abstract and not world-guided. The former are thick, the latter are thin.[4]

4 *Cultural specificity*: While Walzer (1994) explicitly picks up on the use of the term by Geertz (1973), he uses the contrast primarily to refer to the cultural specificity of moral terms, where the thick refers to what is specific to a local place and historical context, and thin to what is claimed to be universal. The assumption appears to be made that the cultural specificity and interpretative depth come together, that the anthropological project of uncovering layers of depth in understanding will take one at the same time to that which is particular to place and time. Walzer in introducing the notion of thick descriptions refers to Geertz's usage, that of interpretative depth, but then shifts immediately to the notion of cultural locality (Walzer, 1994, p. xi).

These different senses of the terms 'thick' and 'thin' are distinct and draw different contrasts. What relations do they stand to each other? There is a prima facie case for some close relation between the first and second senses, for the claim that a thick theory of the good will use terms that involve greater interpretative depth than those of a thin description. A similar case might also be made for the relation between the first and third senses. The relationship between the second and third is more complex. It will be the case that a thick theory of the good will be written a normative vocabulary that is thick in Williams' sense. However, some of the primary goods that are specified by a Rawlsian thin theory of the good will also be thick concepts in Williams' sense – consider for example the concept of 'self-respect'. However, in the context of the debates around thickness and moral universalism what is of particular significance is the relation of thickness in the last sense, to mark degrees of cultural specificity, to the first and second uses of the term: to mark differences in interpretative depth and to mark differences in conceptions of the good. Walzer, as we noted, assumes that there is a direct relationship between them. However, this strikes me as too quick. There is more to be said about different senses of cultural specificity.

We need to distinguish between a good that is a specific to a local culture and a local cultural specification of a good. The claim that goods are specific to local cultures is the one that raises the spectre of relativism; what is a moral good in one culture is not in another. To talk of a local cultural specification of a good is different. The good can be a universal one – it is the particular specification of the good that is local. It is the assumption that thickening descriptions in the Rylean sense of adding interpretative depth take us to goods that are specific to a local culture that appears to ground the view that the anthropological project moves in opposite direction to the more universalising aspirations in ethical conversation. However, that assumption needs to be questioned. Rather engaging in the venture of thick descriptions can take us rather to local specifications of some shared good.

The point can be illustrated by an early exercise in cross-cultural interchange that we have already considered in another regard chapter one of this book. The existence of a plurality of different ways in which human life is lead and the problem this poses for claims to universality are not new. So just as Aristotle appeals to the kinship of humans experienced in their travels (Aristotle, 1985, 1155a 21–22), so other classical Greek texts had pointed to the strangeness experienced. Thus the passage recounted by the greater teller of strange tales for distant lands, Herodotus, quoted earlier in Chapter 1:

> When Darius was king of the Persian empire, he summoned the Greeks who were at his court and asked them how much money it would take for them to eat the corpses of their fathers. They responded they would not do it for any price. Afterwards, Darius summoned some Indians called Kallatiai who do eat their parents and asked in the presence of the Greeks . . . for what price they would agree to cremate their dead fathers. They cried out loudly and told him to keep still. That is what customs are, and I think Pindar was right when he wrote that custom (nomos) is king of all.
>
> (Herodotus, *Histories*, 3.38 translation from McKirahan, 1994, p. 391)

Herodotus in this passage is often taken here to be illustrating the variability of standards of conduct according to the customs of a culture against those who held that there existed universal standard grounded in nature (physis). What is horrific for the Greek is the norm for the Kallatiai and vice versa. Custom is king of all.

However, this appeal to variability in responses to found a thorough going relativism is less convincing than first looks suggest. The passage illustrates not radical differences in the basic goods of two different cultures, but radically different specifications of a shared good. Darius' elicitation of protests relies on understandings that are shared by the two groups, the Greeks and the Kallatiai, about the existence and value of special ties to particular others, more specifically to one's dead parents. The two groups express those relationships in different ways, one by cremation, the other by eating, but that the differences make sense against the background of a shared understanding of the respectful treatment of the dead. One part of cross-cultural interchange, where this can issue in understanding, lies precisely in the interpretation of the actions of others that makes them not only understandable but expressive of relations and attitudes that we share at a deeper level. What in one culture would be shocking, eating your mum and dad, in another turns out to be expressive of a relation of respect to them. The story of Darius would be very different if, as we tried to unpack the act of eating one's parents, it turned out to express an attitude of deep hatred between generations.

Against the standard view that the thick moves us to merely local goods and moral terms, it is often, although not always, the other way around – that it is as one moves to descriptions with greater interpretative depth that values, practices and goods that are shared emerge. In the Darius passage, it is as one thickens the description – from the physical description of the act as eating one's parents to the thicker description of an act of expressing respect for one's parents – that what is shared across the cultures becomes visible. It is at the thin level of description that radical differences are apparent. Similar points apply to examples discussed in Chapter 7. It is as one moves from thinner to thicker descriptions – from 'burning a forest' to 'clearing the land' to 'maintaining the agricultural land of a family' – that common understandings and goods begin to emerge. The thin level of description is likely to occasion a judgement that the act is one of mere destructiveness. It is as specific meanings and goods are uncovered, for example the very concrete and particular ways land embodies the life of a community that shared goods are made apparent. That interpretative depth often takes us to what is shared is not an accident. Shared practices and values are a condition of common action and communication. This is not to deny the possibility of disagreement or of social and moral practices that are radically different. However, disagreement and dialogue are possible only against a shared background of understandings (cf. Midgley, 1981, pp. 72–73).

The passages quoted in Chapter 7 from peasants and farmers speak not just to a local audience, but a wider potential audience. And it is in virtue of this fact that they should be an occasion for critical reflection on nature conservation and deep green attitudes to nature. Thus one effect of conversations with peasants and farmers of the kind quoted is that through their expression of shared understandings and goods that our attitudes and perceptions of natural are changed. We look on a herb-rich meadow and see not only the flowering of biodiversity, but also the decline of an agricultural community; or we look upon a nature park and see not the protection of nature, but the disruption of a community's lived relationship with its environment through a state artifice. Correspondingly one failure of respect in an intercultural context lies in a blindness to the specific meaning relationships, objects and places have to others. What is absent in the case of the wilderness model is the acknowledgement that these are the homes of others, not a wilderness. However strange and alien the non-temperate regions might be to a European like Huxley, that is not what it is like for the indigenous.

There is, then, no incompatibility between a universal ethical reflection and the project of uncovering interpretative depth. Where an incompatibility does exist, it is with a particular style of philosophical reflection on values of which Kantian and utilitarian ethical systems are typical expressions, as are the forms of ethical proceduralism that divorce the norms of engagement in argument from substantive ethical claims. This style of reflection abstracts from the thick vocabulary in which most particular judgements are expressed for a thinner language of 'the right' and 'the good'. That language is, as I noted above, taken to be a condition of the possibility of standing outside our own ethical practices and formulating sceptical questions about them. We must be able to ask the question 'But is x good?', for example, 'But is humility before nature good?' The claims are mistaken. It is not the case that the use of thick concepts rules out theoretical reflection and general principle in ethics. To ask 'Is humility before nature good?' is not to ask whether it has some property of goodness or to ask for the expression of some preference, but to raise questions about the relationship of such humility to other particular evaluative claims we might make – for example, about its compatibility with other admirable human accomplishments. It is to place an ethical concept within the critical company of other thick evaluative concepts. It is particular thick concepts that have reason-giving powers, not more abstract thinner concepts.

To make these points questioning the assumption that thickening descriptions of the social practices and goods entails moving towards that which is purely culturally local provides at least the starting point for a defence of the kind of Aristotelian conception of public life that some recent versions of liberalism have been concerned to reject. As we noted at the start of section 8.1, at least one of the liberal grounds for a minimal ethical vocabulary in political life is the thought that any thick conception of the good will be inconsistent with the pluralism typical of modern life. Given the plurality of ways in which a good life can be led in the modern world, justifications of public policy that appeal to a particular thick theory of the good or a thick normative framework, necessarily leads to the imposition of a particular local contested conception of the good. Public justifications of policy must only appeal to a thin ethical vocabulary. The arguments of this section provide the starting point to a reply to those objections to an Aristotelian politics. There is considerable role for plurality from within an Aristotelian conception of the good.[5] Shared goods are open to very different specifications. In particular, it is possible from within an Aristotelian perspective to both defend the claim that a relationship with nature as a constitutive component of the good life for humans while allowing a plurality in the ways this relationship can be realised.

Nussbaum (1990) makes similar points in noting that any defensible theory of the human good should be thick but 'vague in a good sense': it will allow that are a variety of different specifications of ways in which humans can lead a flourishing life, while offering an account of the powers, capacities and needs that make us the kind of being we are which does delimit the target of what a flourishing life can be (Nussbaum, 1990, p. 217 and passim). The difficulty that a universalist conception of human flourishing has lies in steering a course between vacuity on the one hand, and an implausibly narrow specification of flourishing on the other (Williams, 1995b, pp. 199–201). However, that is a difficulty not an impossibility: a course can be steered and as such it still has real work to do. For example, to take a standard and I think true claim that is found in naturalistic theories of the human good that go back to Aristotle, as humans we are beings that need intimate relations to particular others. That claim does not determine some particular form such relations have to take – it allows of variability. The relations can take a variety of specific forms in different social and cultural settings. The Darius story illustrates just how dramatic that variability can be: however, it leaves room for criticism of a society like our own, in which those who are old and without wealth are often excluded from ties of affiliation with others. Similar points apply to nature. One can, as Soper puts it in summarising Raymond Williams, 'recognise both the permanence of the need for the countryside and the cultural relativity of its expression' (Soper, 1996, p. 214). In recognising the former gives one critical ammunition against social forms that deny the satisfaction of that need; in recognising the latter one can accept that different kinds of landscape can satisfy that need.

8.3 Plural goods and social union

The defence of an Aristotelian perspective in section 8.2 is however only partial. The conflicts noted in Chapter 7 and in section 8.1 do not merely point to the existence of the ways in which there exist different cultural specifications of the goods of a human life. The conflicts between farmers and conservationists in the Yorkshire Dales, for example, would not simply disappear on the specific goods in question being properly understood. They exist also in virtue of the plurality of goods themselves and of real conflicts between those goods in particular conditions. The plurality is evident here in the way in which the same environmental objects and states are evaluated differently under different descriptions – as a particular kind of ecosystem, as a wetland of a certain variety, as a feeding ground for certain migratory birds, as a particular habitat, as a landscape, as a soil type suitable for a specific kind of agriculture, as a place that is inhabited by some community, as an industrial wasteland, and so on. It is under such descriptions that objects are evaluated. A location can be at one and the same time a 'good A' and a 'bad B'. A location, say an old quarry, may have considerable worth as a place – it may embody in a particularly powerful way the work of a community – but little worth as a habitat, an ecological system or as a landscape. A wetland might be of considerable worth as a habitat and as an ecosystem, but at the same time have little value as a landscape or as agricultural land.

These are conflicts between values, not merely between different beliefs about what is of value.[6] However, the values are embodied within different social practices and ways of life. Different values are combined into particular ways of life with their specific virtues. Being a hill farmer in the Yorkshire Dales is a way of life with its particular virtues that involves care for the land and maintenance of its productivity in harsh conditions.[7] Being a hill farmer is not the same as being a park-keeper, a point that farmers often stress.[8] And while habitats that now have conservation orders placed upon them may be the result of particular patterns of pastoral and agricultural activity, they were rarely the aim of those activities: skylarks and meadow herbs were the unintended beneficiaries of farming practices, not their direct purpose. The farmer looks at a conserved landscape and sees from within the practice of farming 'a mess'. A conservationist looks at well-maintained agricultural land and sees 'an agricultural desert'.[9] The first order conflicts of values find expression in conflicts between participants of different practices and ways of life.

The point here is one that is brought out in a very different way in Auden's stanzas and the case it raises against an environmental virtues ethic. Against those who insist on the need for appreciation of 'the wonder and beauty of the natural world', surely it might be argued there are those who, like the Auden presented in the 'Letter to Lord Byron', live flourishing lives with little or no contact with the nature – for whom the need for the countryside is not felt. Within our own culture, there are individuals who live perfectly good flourishing lives not just below the snowline, but with little or no contact with nature, who love the life of the city, dusty library rooms, art galleries, conversations in restaurants and only with some reluctance can be goaded into the occasional excursion into the garden or the city park, and does so merely to humour their outdoor companions. It is a form of life I might not share but one I have met and can appreciate. There are different forms that a flourishing life can be led which realise quite different kinds of good.

However, these observations are compatible with the inclusivist version of the Aristotelian position, which accepts that the goods that go up to make a flourishing life are irreducibly plural. A flourishing human life contains a variety of intrinsic goods which cannot be reduced one to another (Aristotle, 1985, 1096b 23ff.). Given that plurality and the fact that we are not angels but beings that work within limits of capacity, time and resources, we face practical choices in their pursuit of such goods. Some groups of activity will make up a form of life that is incompatible with others: one may not be able to lead a life of contemplation and a life of action. The pursuit of such activities may conflict with the calls of other relationships. Where choices exist, such conflicts give rise to the practical dilemmas of individual lives. There is no algorithmic procedure in such cases – rational choice is made on the basis of practical judgement. Choices may be more or less difficult. However, the limitations of individual lives force such choices upon us. And we can expect different individuals to realise different goods in their lives. The Audens may lead culturally fulfilling lives in one way, others lead it in another. The hill farmer realises particular human goods that are unavailable to those in the city.

What is true, however, is that a good society will be one in which there is expression of care for land and where appreciation for the wonder and beauty of nature in the lives of some of its members, alongside those who pursue other goods.

The recognition of this point provides the foundation however of Aristotle's defence of plurality within the polis against Plato. The goods realised by the polis, while bounded, are wider than those any individual member can realise. I cannot realise excellence in several fields – we can. This greater range of the goods realisable within the polis enhances the lives of its members. Through my relations with others, I can have a vicarious interest in these goods. Consider the case of friendship: to have friends with a diversity of interests and pursuits extends me. In caring for the good of my friends, I care for the success of the projects in which they are involved, for their realisation of excellence in the activities they pursue. A friend 'shares his friend's distress and enjoyment' (Aristotle, 1985, 1166a 8). Thus in friendship the ends of another become one's own. Hence, while I may not be involved in such activities, I have a vicarious interest in the achievement of goods within them. My concerns are extended by those around me. While there are limits to the goods I can personally achieve, I can retain an interest in their achievement through others for whom I care. Hence, given relations of civic friendliness of a kind Aristotle assumes in an ideal polis, a community in which the largest number of goods can be realised will enrich the lives of all its members. I have little knowledge nor appreciation for opera or the visual arts, but it is better for me that I live among those who do, just as it is better for the Audens that they live in a society where others do appreciate the goods of life above the snowline.

It is in virtue of these points that Aristotle holds that humans can achieve a complete and self-sufficient good only within the polis in which individuals are able to enter a variety of relationships and pursue diverse and distinct goods. The pursuit of these particular goods will be itself a social enterprise that will take place within different associations. The end of the polis is not some completely separate good over and above these partial goods: its end is rather an inclusive end. Hence Aristotle's characterisation of the polis in the opening paragraph of the *Politics* as an association of associations (Aristotle, 1948, 1252a 1–7). The good that the polis pursues is an inclusive good: it contains all those goods sought in more particular associations. The polis has the comprehensive goal of realising the good of the 'whole of life'. On this view, the polis does not replace other partial associations, but is rather a community of communities containing a variety of associations realising particular ends (Aristotle, 1985, 1160a 8–30). It has the architectonic function of bringing order to and resolving conflicts between these goods, just as practical reason brings order to the pursuit of the variety of goods pursued by an individual through a lifetime. The polis is plural in the relationships, goods, and associations that it contains. To view the purpose of politics as the pursuit of the good life is compatible with a pluralist view of the political community.

Where the Aristotelian conception differs from certain versions of the modern liberal conception of politics is in the account of the proper relations between different associations should be. On at least some modern versions of liberal neutrality – and I should add I do not think that Rawlsian liberalism belongs to them (section 79 of *A Theory of Justice* contains an excellent articulation of the Aristotelian ideal of social union)[10] – liberalism is founded upon a pluralism of indifference. In our community we do this, in yours you do that and as long as we do not interfere with each other in the pursuit of other ends, all is well; I as an individual have this ideal, you have another, and the liberal order allows us to proceed without interfering with each other. The Aristotelian position is based rather on a pluralism of recognition – that a community in which different goods are realised extends the interests of each. Such a pluralism of recognition applies not only to the recognition of the different goods that different communities and associations pursue, but also to the different specifications they offer of the good life. In both global and local multicultural context one can not only recognise that different communities express relations and attitudes we share at a deeper level in different ways, but also find the different modes of expression enriching: the differences capture modes of relating to each other and nature that extend our understanding of such relations. At the same time, pluralism of recognition does not entail that we give up our critical faculties in the face of others and ourselves (cf. Taylor, 1992, pp. 61–73). The obverse of recognition is criticism. To recognise the virtues of an association or community is at the same time to recognise the possibility of vices, say in purely exploitative relationships to both humans and non-human nature. Nor does pluralism of recognition rule out the existence of real conflicts between values in particular situations and of social choices, sometimes painful, between them. Such conflicts can be expressed as conflicts as ways of life themselves. However, if there is loss, say of a landscape that embodies the life of a community, there needs to be recognition of what is being lost and expression of appropriate responses to that loss.

8.4 Narrative

This point about the multiple ways in which a good life can be lived and the plurality of goods it can contain still, however, does not I think capture the extent of variability in which a flourishing life can be led. The Aristotelian position is often associated with an 'objective list' account of the content of well-being. A list of goods is offered that correspond to different features of our human needs and powers – personal relations, physical health, autonomy, knowledge of the world, aesthetic experience, accomplishment and achievement, a well-constituted relation with the non-human world, sensual pleasures and so on. Increasing welfare then presented is a question of maximising the score on different items on the list or at least of meeting some 'satisficing' score on each (Simon, 1972). This picture of welfare is compatible with the kind of variability I have just outlined. The different items on the list can be satisfied in a variety of different ways, and the descriptions on the list must be such that they allow for different instantiations of the goods. However, while the appeal to such general values is in part right, the maximising or satisficing approach it suggests is unsatisfactory.

The appeal to an objective list misses the role of history and narrative in appraising how well a person's life goes, which has been discussed in detail in Part II of this book. The importance of temporal order is already there at the biological level. In appraising the health of some organism, the path of growth and development matters, not just some static scoring system for the capacities it has. Once we consider the cultural and social dimensions of human life this temporal order has a stronger narrative dimension. Answering specific question of how a person's life can be improved is never just one of how one can optimise the score on this or that dimension of the good, but how best to continue the narrative of a life. The question is 'given my history, or our history, what is the appropriate trajectory into the future?' This is not to say there is just one possible trajectory, nor that there is any algorithm to determine it. Our history constrains, it does not determine. However, it moves us away from maximising concepts and entails a more radical variability. It may be that from some static maximising perspective, the best course for a person would be to abandon the isolated farm on a fell that he has farmed for the past 60 years and which was farmed by his family before him, and to move to a retirement home. His material, social and cultural life might all improve. But given the way his life has been bound up with that place, that would involve a disruption to the story of his life. The desire to stay 'despite' all the improvements offered is a quite rational one. Likewise, for communities with their particular traditions and histories, what may be 'maximising' from an atemporal perspective, may not for that particular community.

As I have already argued at length in Part II of this book, this historical and narratory dimension of our individual and communal lives is of particular significance in the environmental sphere. We enter worlds which are rich with past natural and human histories, the narratives of lives and communities from which our own lives take significance. The problem of conservation is, or should be construed as, the problem of how best to continue the narrative; and the question we should ask is 'What would make the most appropriate trajectory from what has gone before?' (Holland and Rawles, 1994, p. 37). It is because it is about the negotiation of a narrative order between past and future, that it is not simply about 'preserving the past'. Indeed once the significance of the role of a narrative order is understood, we can understand more clearly just what is wrong with that approach. One major problem with the heritage industry is the way it often attempts to freeze historical development. A place then ceases to have a continuing story to tell. The object becomes a mere spectacle, a museum piece, taken outside of any common history. Hence the proper objection from the farmer quoted in Chapter 7: 'National Parks, English Nature, they'll finish up with all the farmers running about in smocks, like museum curators.'

8.5 The good life above and below the snowline

In Chapter 7 and at the start of this chapter I have attempted to uncover some of the more troubling aspects of some recent versions of environmentalism and the potentially authoritarian policies that they might involve. However, I have also suggested in this chapter that the proper response is not to shift the language of public policy to a thin moral minimalism in the manner suggested defenders of liberal neutrality. On the one hand this very moral minimalism brings its own potential problems of silencing marginal voices from public debate. On the other hand, a thick and rich public vocabulary in not necessarily authoritarian or incompatible with the pluralism of the modern world. Neither does it entail a shift to purely local goods and to relativism.

A broadly Aristotelian position which defends environmental goods as constitutive of the good life is consistent with recognition of the plurality of ways our relations with the natural world can be lived. It is compatible with the recognition of distinct cultural expressions of such relations, the existence of forms of life in which the pursuit of other goods render such relations marginal, and with the special place the particular histories have in determining what is the appropriate future trajectory for particular individuals and social groups. The good life can be lived both above and below the snowline. Aristotelian conceptions of public life are quite compatible with pluralism of recognition of the plurality. By the same token they are quite able to provide the basis for robust criticism of the growing authoritarianism of some forms of environmentalism, and in particular the alliance of certain conservation agencies, developing world elites and corporate wealth and power which has led to the appropriation and exclusion of poor and marginal populations in the name of a particular flawed wilderness model of the environment.

However, once the pluralist account of the goods of human life of the kind outlined in this chapter is granted, as we noted above the existence of conflicts between those different goods has also to be recognised. Correspondingly, the question of how such conflicts are to be lived with remains. I say 'lived with' rather than 'resolved' since there may be no possibility of resolution. The assumption that they do require resolution is one that runs through much of the literature. One of the promises of the market-based approaches to environmental policy making discussed in Part I is that it can offer a set of calculative procedures for resolving such conflicts employing monetary prices. In Parts I and II I have argued that this approach should be rejected and outlined deliberative alternatives to those particular monetary approaches. Deliberative models are themselves often presented as a distinct mode for resolving such normative disputes. However, the standard deliberative alternatives have their own difficulties. In the final part of this book I consider the strengths and limitations of those deliberative approaches.[11]

Part IV

Deliberation and its discontents

9 Deliberation, power and voice

In Part I of this book I criticised market-based approaches to environmental policy making and provisionally defended a deliberative alternative. In the final chapters I want to look more critically at this deliberative response. Deliberative or discursive models of democracy have enjoyed a justifiable revival in both theory and in practice. Against the economic picture of democracy as a surrogate market procedure for aggregating and effectively meeting the given preferences of individuals, the deliberative theorist offers a model of democracy as a forum through which judgements and preferences are transformed through reasoned dialogue between free and equal citizens (Elster, 1986, 1998c; Manin, 1987; J. Cohen 1989; Dryzek, 1990, 2000; Fishkin, 1991; Miller, 1992; Bohman, 1996; Bohman and Rehg, 1997; Guttman and Thompson, 1996; Habermas, 1996; Rawls, 1996; Chambers, 1997; Sunstein, 1997; Smith, 2003). The main, although not only, theoretical source of recent deliberative theory has been Kantian. Deliberative politics embodies the Kantian ideal of the public use of reason. In liberal theory, the work of Rawls (1996) has been particularly significant in the development of a Kantian deliberative politics. The Habermasian account of communicative rationality which runs through much of the recent literature on deliberative democracy similarly has its roots in Kant's defence of the public use of reason as a condition of enlightenment: dialogue is rational to the extent it is free from the exercise of power and strategic action, such that the judgements of participants converge only under the authority of the good argument – 'no force except that of the better argument is exercised' (Habermas, 1975, p. 108).

The recent revival of deliberative democracy has been expressed in practice in two distinct forms. First, formal policy practice has seen the development of a variety of 'new' formal deliberative institutions which have been introduced alongside 'older' democratic institutions and which are often presented as experiments in deliberative democracy. These include citizens' juries (Armour, 1995; Aldred and Jacobs, 1997, Coote and Leneghan, 1997), citizens' panels (Rippe and Schaber, 1999), consensus conferences (Joss and Durant, 1995), mediation panels (Weidner, 1993), focus groups (Macnaghten et al., 1995), in-depth discussion groups (Burgess et al., 1988) and round tables (Gordon, 1994). At the less formal level, the growth of associations and movements within civil society has been read, in particular within the Habermasian tradition, as a development of a discursive politics that exists between the state and economy.[1] There are at least two aspects to this deliberative reading of civil society.[2] First, social movements themselves are ongoing conversations which sustain particular arguments and identities. Those conversations can take place both in public fora, for example, the various counter-fora to organisations such as the WTO and World Bank, and more widely and often less visibly through local and global networks of activists. Second, movements are also the medium through which public arguments can develop by extending the topics of and resources for wider public debate. The latter points picks up on the theme in associational theories of democracy – that it is only through intermediate associations in public life that citizens' voices can be effectively articulated and heard.

Both the theory and practice of deliberative democracy have been particularly developed in the environmental sphere (Dryzek, 1992, 2000, ch. 6; Jacobs, 1997; Eckersley, 1999). At the level of civil society, social movements and citizens' initiative responding to a series of environmental problems from environmental degradation to the unequal distribution of environmental goods and harms at both local and global levels have been central to the renewal of civil society. They have been particularly central in the development of deliberative spheres at the global level that can challenge the neo-liberal project of globalisation. At the formal level, environmental policy has been a prominent site for the development of new experiments in deliberative democracy like citizens' juries, focus groups and consensus conferences. In this context the experiments have had particular power as a response to the reason-blindness and distributional failings of traditional market-based approaches to environmental decision making discussed in the first part of this book.

Cost-benefit analysis as an approach to the resolution of value conflicts about the environment is reason blind. The strength and weaknesses of the intensity of a preference as measured by a person's willingness to pay at the margin for their satisfaction do count in a decision; the strength and weakness of the reasons for a preference do not. Preferences are treated as expressions of mere taste to be priced and weighed one with the other. It offers conflict resolution and policy without rational assessment and debate. Politics becomes a surrogate market which completes by bureaucratic means what, within neoclassical theory, the ideal market is supposed to do – aggregate efficiently given preferences (Sagoff, 1988; O'Neill, 1993a). However, since environmental conflicts are open to reasoned debate and judgement which aim to change preferences not record them, it follows that different institutional forms are required for their resolution. Since conflict is open to reasoned adjudication, discursive institutions are the appropriate form for conflict resolution. The forum, not the market, becomes the proper institutional form. The central point of the distributive criticism of cost-benefit analysis is not simply that it focuses on issues of efficiency rather than of distribution, but as we noted in Chapter 3 that efficiency and distribution are within the theory not independent. If property rights are changed, so also is what is efficient. Different initial distributions entail differences in whose preferences are to count. Hence, environmental policy and resource decision making cannot avoid making value choices which include questions of resource distribution and property rights (Samuels, 1972;

Schmid, 1978; O'Connor and Muir, 1995; Martinez-Alier et al. 1999). In practice, cost-benefit analysis tends to conservatism by assuming a status quo in the distribution of effective rights. Given raw willingness to pay scores, the preferences of the poor will count for less than the well-off, the interests of future generations and non-humans will count at best precariously. Deliberative institutions are often presented as responses to these distributive failings. The distribution of resources and property rights is not presupposed as it is in market methods, but can itself be an object of public deliberation. Thus goes the case for deliberation and against market-based approaches to policy making.

However, at the same time the appeal to the language of deliberation and deliberative democracy has been increasingly taken up by state and corporate actors, often to a very different effect. Experiments in deliberation, like citizens' juries and focus groups, have been adopted by state actors as part of a response to what is seen as a 'democratic deficit' in existing institutions. That deficit is taken to be expressed on the one hand in the decline of participation in formal electoral processes, indicated by the low turnouts in elections, and in the increasing dissatisfaction and distrust in elected representatives and on the other in the perceived correlative rise in the extent of and public sympathy with direct action. The appeal to deliberative models has been used both to respond to both problems. Experiments in deliberation are presented as responses to the problem of 'democratic deficit' by forming part of project to 'democratise democracy'.[3] At the same time the language of deliberation is increasingly used to counter direct action in two related ways: first, in the attempt to engage NGOs, associations and activists in stakeholder fora and other 'inclusionary' and 'deliberative' bodies with corporate actors, state agencies and international bodies such as the World Bank and World Trade Organization; second, the employment of the language of deliberation and democracy back against the direct action movements who refuse to engage but rather sustain an oppositional stance. In particular, as direct action movements against globalisation have grown there has been an increasing use of the language of deliberation and free discussion by bodies like the WTO and the World Bank against the 'disruptive' protest of the movements against neo-liberal globalisation and their refusal to engage in 'constructive dialogue'.[4]

These responses by state and corporate actors highlight some of the major problems with the appeal to deliberative models of democracy where these abstract from the large asymmetries of power and resources against which deliberation takes place. The ideal of deliberative democracy is typically characterised in terms of dialogue between formally and substantively equal participants. A rough equality of power of voice is a condition for proper deliberation. Rationally acceptable claims are those that can be redeemed in conditions of equality in voice. In the absence of those conditions deliberative models of democracy have their own problems of legitimacy.

Consider recent formal experiments in deliberative democracy. Within deliberative fora voice can be unevenly distributed. The capacity and confidence to speak and, more significantly to be heard, differs across class, gender and ethnicity. Just as 'willingness to pay' in cost-benefit analysis is unevenly distributed, picking up inequalities of income, so also is the capacity or 'willingness to say' in more formal settings of deliberative fora, and, more importantly, the willingness and capacity to be heard. Many of the standard observations about the micro-politics of conversation play out in recent experiments in deliberative democracy (Fairclough, 1999). Hence, there is power in the response to the deliberative critics of economic valuation: that the deliberative alternative is a way in which those with the power of voice are able to exercise it over those who lack voice (Beckerman and Pasak, 1997). It is not just the internal workings of deliberative institutions that matters here but the context in which they operate.

Recent experiments in deliberative democracy contrast with traditional representative democracy on the one hand and with the formal and informal spaces of public deliberation in civil society on the other. In contrast to those ongoing public conversations, recent experiments in deliberation have, from the perspective of participants, certain features: the space in which deliberation takes place, who is included, the agenda on which participants are brought together, the opening and closure of conversation, the identities which participants are ascribed, the afterlife of the results of deliberation and their effects on policy making, are normally organised by others. While many of these features depart from standard accounts of deliberative ideals,[5] their presence need not as such undermine their legitimacy. Whether they do depends on the context in which deliberation takes place. Moreover, these features can be subverted by the process of deliberation.[6] However, they do leave deliberative institutions open to being used strategically by powerful actors in particular ways. For example, some of the new institutions that are claimed to be deliberative, in particular focus groups, are often employed in political practice not to allow deliberation to take place but rather to close it down. They are used to gather information of likely responses to different potential policies and actions not to open up debate but to anticipate and forestall it. The origin of the focus group technique in market research is not without its implications.[7] Where deliberation is public, power can be exercised in the framing of issues prior to discussion and in the choice of the constituency for debate. For example international companies have used processes of stakeholder engagement involving sophisticated techniques to disaggregate the different actors in communities, to deal with representatives of those who might be allies and to use local alliances around environmentally damaging activities against larger environmental regulation (Cochrane 1998).

If effectively captured by powerful institutions deliberative processes potentially provide powerful legitimation tools – indeed much more powerful than expert-based techniques like cost-benefit analysis, which are widely perceived to lack democratic legitimacy. Correspondingly there are contexts in which dialogue may simply be a means through which those who lack social power concede to those who do not. Hence, where there are large asymmetries of power, for those without power dialogue might not be the appropriate response. Given the asymmetries of power between the actors, and the control of the conditions and topics of dialogue by those with power, it is entirely reasonable from within a deliberative account to refuse to engage in dialogue. The conditions required for reasonable dialogue are absent. Indeed, there are conditions in which to refuse to engage in speech is itself a form of communicative action. To engage in dialogue with another involves mutual recognition of partners in the dialogue as legitimate actors. For that reason, one way of expressing the denial of legitimacy to an actor is to refuse enter dialogue.[8]

Finally, any formal process of inclusion of parties within deliberative fora has to be considered against the effect of this on wider public deliberation. Social movements themselves are partly ongoing deliberative conversations. The central argument for associational models of democracy, that it is through a variety of intermediate associations in social life that individual voices are best articulated and heard in public conversations, does offer a strong argument for such wider conversations. Engagement with state, corporate and transnational bodies can affect the quality of deliberation within social movements (Dryzek, 2000, ch. 4). It can disrupt the relations within movements and hence the unconstrained and relatively egalitarian forms of dialogue that some sustain. And it can shift the nature of the deliberation by reframing the terms of discussion in ways that can be accommodated within the dominant discourse of free markets and globalisation, rather than sustaining debate that is oppositional to that order. Civil disobedience involving illegal action is itself a public expressive act that opens up new spaces for public dialogue (cf. Cohen and Arato, 1992, ch. 11). Theatrical and 'disruptive' acts of protests even where they do not constitute moves in an argument, have themselves opened up topics for wider public debate that were often off-limits in official public discourse. In the context of narrow neo-liberal agenda of official discourse on globalisation, the carnivals of protest have been a powerful and legitimate contribution to public deliberation.

Now, it might be rightly objected that these criticisms of the recent use of the language and rhetoric of deliberation from state and corporate actors are not criticisms of deliberative democracy as such. Indeed they are for the most part internal criticisms that proceed from the assumptions of deliberative theories. It is an internal feature of most deliberative models that deliberative institutions fail to the extent that such asymmetries of power and strategic action are present in determining the outcomes of dialogue. Now that still raises real problems of how far any approximation to deliberative institutions is possible in existing or alternative social, economic and political conditions, but that is as Habermas notes about 'the possibility of effectively institutionalizing rational discourse' (Habermas, 1993, p. 57) and not about the account of rational discourse that deliberative democracy is taken to offer. The deliberative theorist can then claim that the potential problems with the practice of new deliberative institutions do not undermine the deliberative theory of democracy as such. Rather deliberative theory provides the normative basis from which an immanent criticism of the failings of putatively deliberative institutions can proceed. As such, the theory can be still said to have purchase. Hence Habermas's claim that while 'the actual course of the debates deviates from the ideal procedures of deliberative politics . . . presuppositions of rational discourse have a steering effect on the course of the debates' (Habermas, 1996, p. 340).[9]

However, while there is much that is right in this internal response by deliberative theorists to problems of power and voice raised by putative attempts to develop deliberative institutions, the problems also indicate important problems with some normative accounts of good public deliberation. In the following chapters I will examine in detail three sets of limitations with a number of recent theories of deliberative democracy. The first concerns the nature of deliberation itself. Any deliberative criticisms of existing experiments in deliberation rely on some account of what counts as good deliberation which is itself open to debate. Habermasian and Rawlsian accounts of deliberative democracy appeal to a broadly Kantian model of deliberation that is particularly questionable in the intellectualist account of deliberation it presupposes. The Kantian model is not just problematic in virtue of the range of forms of communication that it precludes from its ideal account of deliberation and hence the voices it potentially excludes.[10] It is also fails in its account of what good public deliberation is like. In particular two features of deliberation that have been central to the rhetorical account of deliberation that stems from Aristotle are badly dealt with in the Kantian tradition: the role of testimony and the appeal to the emotions. Neither can nor ought to be eliminated from public deliberation. In particular, in the environmental sphere where scientific expertise is called upon for the identification and assessment of risk, many of the central questions of deliberation are not a matter of directly assessing evidence and argument, but more a matter of ascertaining whose testimony is credible and trustworthy – of deciding *who* to believe in what institutional conditions. In Chapter 10 I will develop this Aristotelian account of public deliberation in more detail. I will argue that while Kantians are right to be critical of some rhetorical models of public deliberation, this Aristotelian account of deliberation does render it consistent with the central enlightenment values that the Kantian theorists aim to defend.

The second set of problems I will address in Chapter 11 concerns issues around the legitimacy of representation for deliberative models of democracy. The questions of whose voice is present at deliberation, under what representation and with what forms of accountability run through recent problems of deliberative democracy, both in the development of new experiments in deliberation and in the wider deliberations of civil society. First, with respect to civil society, the deliberative understanding and defence of civil society movements outlined above does not entail that they do not have their own problems of representative legitimacy. The problems of asymmetries of power and voice clearly exist among the different associations of civil society and indeed between these and those who are marginalised and have no vehicles for public voice.[11] And as many movements are increasingly recognising in reflections on their own practice, there exist large problems about the representative legitimacy of different associations, especially concerning the existence and nature of mechanisms of accountability to those for whom the associations claim to speak. This is a particular problem in environmental organisations who claim to speak for future generations and non-humans. Representation of future generations and non-humans raises well-rehearsed difficulties for deliberative theories of democracy (Goodin, 1996; Eckersley, 1999). Their direct voice is necessarily absent, and hence, as with market methods, consideration of their interests relies upon indirect representation. The absence of direct representation raises major problems concerning the ethical and political legitimacy of decisions made in the absence of their voice. However, at the same time it raises problems also for forms of environmental action and advocacy that are legitimised by appeal to the claim that protagonists are speaking and acting on behalf of those who are without voice. Second, many recent experiments in deliberative institutions – citizens' juries, in-depth discussion groups, consensus conferences and the like – are small and hence raise issues about the source of their legitimacy as representative of larger populations.[12] The problem here lies not with the issue of statistical representativeness that is often raised, but rather with concerns around their representative legitimacy. Whose voice is present, under what description and under what conditions of legitimation? It is the main contours of answers to those questions which will concern me in Chapter 11.

Chapter 12 considers these issues of power and voice in more detail. Deliberative criticisms of existing deliberative institutions in the end pushes beyond the purely political and cultural models of deliberative institutions that are defended in much of the literature both within liberal theory and increasingly from the tradition of critical theory. They point in the direction of traditional egalitarian claims about the structural conditions for democracy: formal equality in democratic rights is insufficient without a rough equality in social and economic power. Standard accounts of deliberative democracy recognise that deliberation requires both formal and substantive equality. The latter is taken to be present if internal dialogue is unaffected by external distributions of power and resources: 'The participants are substantially equal in that the existing distribution of power and resources does not shape their chances to contribute to deliberation, nor does that distribution play authoritative role in their deliberation' (J. Cohen, 1989, p. 23). The liberal ideal is of autonomous deliberative institutions protected from those patterns of power in the economy.[13] However, where large inequalities in economic and social power exist, that liberal ideal on its own terms becomes untenable. More generally there is a tendency in much recent work on deliberative democracy to stay at the purely discursive level. The result is a purely symbolic or cultural politics which fails to address the ways in which the structural imperatives of markets place constraints on the actual decisions of actors (O'Neill, 1998a, ch. 1). In Chapter 12 I will consider some of these issues in more detail. I will do so by way of a consideration of the forgotten role that environmental problems played in the socialist calculation debates in the twentieth century.[14] I will attempt to show that the arguments about the environmental limits of market solutions to environmental problems, the central theme of this book, still have a great deal of relevance to current debates about the institutional and structural conditions for an environmentally rational society.

10 The rhetoric of deliberation

The main theoretical source of recent work on deliberative democracy has been Kantian. The Habermasian account of communicative rationality which runs through much of the recent literature on deliberative democracy has its roots in Kant's defence of the public use of reason as a condition of enlightenment (McCarthy, 1995). The work of Rawls (1996) on deliberative democracy shares those Kantian roots. And it is to this Kantian model that appeal is normally made in the design and appraisal of experiments in deliberative democracy. There is (as I noted in Chapter 9) a gap between the theory of deliberative democracy and its putative practice. The practice of deliberation takes place in the context of real asymmetries of power and voice, in which there exist large differences in the capacity to speak and be heard and in which putative deliberative processes are open to various forms of strategic manipulation. In this context the Kantian model is taken to provide the normative basis from which an immanent criticism of the failings of putatively deliberative institutions can proceed. The failings of actually existing deliberation are open to criticism from within deliberative theory itself. However, as I noted at the end of Chapter 9, those problems raise real issues about the nature of deliberation itself.

The main opposition to the Kantian model of deliberative politics in both classical and recent literature has been from the rhetorical tradition. And for some good reasons, recent Kantian accounts of deliberation have defined themselves in opposition to rhetorical models of communication. However, while strong rhetorical models of public deliberation which render all communication strategic are indefensible, there are features of reasoned deliberation that the rhetorical models highlight which point to problems with the Kantian accounts of public deliberation. In this chapter I defend an Aristotelian account of rhetoric and show that it offers a more defensible normative account of public deliberation.

10.1 Reason, rhetoric and deliberation

Recent theories of deliberative democracy inherit the ideal of the free public use of reason which Kant takes to define the enlightenment project. The 'freedom to make public use of one's reason in all matters' is a condition for the emergence of maturity – the capacity and courage 'to use one's own understanding without the guidance of another' (Kant, 1784b, pp. 54–55). The ideal of maturity is closely related to that of autonomy. The heteronomous character is one who lacks maturity, who is willing to let his own judgement and understanding be guided by others and who lacks the capacity, desire or courage to exercise them for himself: when the clergyman, religious or secular, announces 'Don't argue, believe', he believes. To be autonomous is to have maturity and courage in using one's own understanding and judgement. For Kant it is to be guided by reason: 'For reason has no dictatorial authority; its verdict is always simply the agreement of free citizens, of whom each one must be permitted to express, without let or hindrance, his objections or even his veto' (Kant, 1933, A738/B766). Enlightenment requires institutions that embody free public dialogue.

The extent to which recent deliberative theory is Kantian is often missed by some of its own proponents, in part under the influence of the common and mistaken view of Kant as an individualist, 'more concerned with the isolated subject than with social interaction' (Dryzek, 1990, p. 14). Kant's account of public reason links individual autonomy to social conditions for public debate. Every moral principle must meet 'the formal attribute of publicness' (Kant, 1795, Appendix, II). Against the common misconception of Kant as a rigorist for whom rule-following is the core of morality, it places the development of capacities of judgement at the centre of maturity, where judgement itself is constituted by public deliberation: 'Dogmas and formulas, those mechanical instruments for the rational use (or rather misuse) of his natural endowments, are the ball and chain of his permanent immaturity' (Kant, 1784b, pp. 54–55). The power of judgement is required in the application of any universal rule to particular cases and cannot itself under pain of an infinite regress be understood as the application of a rule.[1] This power of judgement is itself social. It relies on the comparison of judgements with others in order to 'escape the illusion that arises from the ease of mistaking subjective and private conditions for objective ones, an illusion that would have prejudicial influence on the judgement' (Kant, 1987, 293–294). The escape from prejudice is identified with the escape from passive reason, the failure to think for oneself, and hence is a condition of enlightenment. Hence, since the comparison with the judgements of others is required to avoid prejudice, the maxim 'think for oneself' and the maxim 'think from the standpoint of everyone else' are related (Kant, 1987, 294–295; cf. Kant, 1933, A820–821/B848–849).

The Kantian model of enlightenment remains influential on recent work on the foundations of discursive institutions necessary for enlightenment. It is central to the Habermasian conception of rational public life. To engage in reasoned dialogue is to aim not at compromise but on convergence in judgements – 'the agreement of free citizens'. The activity of argument presupposes an ideal of free agreement in judgement founded on good reason: in ideal discourse 'no force except that of the better argument is exercised' (Habermas, 1975, p. 108). As such the deliberative theory of democracy relies for its legitimacy on a distinction between reason and power. It inherits the classical distinction between the impersonal force of argument, and the force of institutional and personal power. With it comes the corresponding distinction between uses of language that aim at convergence through reason and those that are strategic in nature and that aim at using whatever means are available to persuade an audience. The distinction is one that is reworked in Habermas' contrast between communicative and strategic action.

The distinction is an ancient one that goes back to that between dialectic and rhetoric in Plato's criticism of the sophists in the *Gorgias*. Plato starts from the distinction between two forms of persuasion 'one providing conviction without knowledge, the other providing knowledge' (Plato, 1987, 454e). Rhetoric is concerned with producing conviction in its audience without knowledge and without reason. Unlike specific sciences such as mathematics it persuades not by teaching but by flattery. It issues not in learning but mere conviction. It is not properly an art at all, but a mere knack. This contrast between rhetoric and rational discourse is central to the Kantian model of deliberative institutions which belongs to the long anti-rhetorical tradition that can be traced back to Plato. Rhetoric is presented as an art of deceit which, since it aims at persuasions without reason, is inconsistent with respect for the autonomy of the hearer. Thus Kant himself, while he allows rhetoric a place in the arts which make no claim to be making assertions of truth, maintains it has no place in discourses that do make such claims:

> Poetry plays with illusion, which it produces at will, and yet without using illusion to deceive us, for poetry tells us that its pursuit is mere play . . . Oratory [on the other hand], insofar as this is taken to mean the art of persuasion (ars oratoria), i.e. of deceiving by beautiful illusion, rather than excellence of speech (eloquence and style), is a dialectic that borrows from poetry only as much as the speaker needs in order to win over people's minds for his own advantage before they can judge for themselves, and so make their judgement unfree.
>
> (Kant, 1987, section 53, p. 197)

Since it is incompatible with autonomy, rhetoric is not permissible. The assumption that rhetoric and reasoned public deliberation are incompatible is one inherited by modern neo-Kantians like Habermas (1987) and Rawls (1996, p. 220). To use rhetoric is to attempt to win a person's beliefs without the exercise of reason. Hence its incompatibility with the enlightenment project of the mature autonomous agent.

The strong version of the rhetorical attack on deliberative theory denies the distinction between reason and force. The view that all public discussion is just strategic action by other means is as ancient as deliberative democracy itself. For many of the sophists all deliberation is ultimately an art of rhetoric understood as an art of manipulation. The distinction between reasoned persuasion and other forms of compulsion is denied. The classic statement of this view is Gorgias' *Praise of Helen*. Gorgias employs the denial of the distinction to excuse Helen's responsibility even where she was 'persuaded by words (logos)' by Paris to leave for Troy:

> For LOGOS which persuaded, compelled the soul, which it persuaded, both to believe what was said and to approve what was done. Therefore the one who persuaded, since he compelled, is unjust, and the one who was persuaded, since she was compelled by LOGOS, is wrongly blamed.

Significantly, the exoneration of Helen is supported by appeal to the power of persuasion in public deliberation: 'a single (LOGOS) written with art (TECHNE) but not spoken with truth delights and persuades a large crowd' (Gorgias, *Praise of Helen*, Diels-Kranz 82B11 translation, from McKirahan, 1994, pp. 376–377). The Sophist offers to teach the power of persuasion understood as an art of manipulation.[2]

Just as deliberative democracy has enjoyed a revival so also has this sophist alternative. The revival has come from two different directions which superficially look very different: on the one hand from public choice theorists who take politics to be the pursuit of self-interest by other means, and on the other from a variety of rhetorical models of discourse popular in some forms of the sociology of scientific knowledge which reject the possibility of the impersonal force of argument. The two views are not in fact as distinct as might be supposed. As I have argued elsewhere much in recent rhetorical studies in science is in effect an application to science of a public choice model of institutions (O'Neill, 1998a, ch. 12, b, c). More generally both views converge on a picture of the presentation of public argument as an exercise in strategic action aimed at other ends. Given this view, strategic communication within deliberative institutions and the strategic use of such institutions in political practice should not be understood as mere misuses of deliberation that fall short of the standards of proper deliberation. They are all one could expect. They reveal the ways in which truth and objectivity themselves are mere rhetorical ploys that are designed to move an audience. The guises of objectivity, truth and impersonal argument are strategic exercises of power, rhetoric that will not reveal itself for what it is. All discourse is an exercise in power. All the analyst should do is to unmask it for what it is. This kind of deflationary rhetorical perspective has been aimed principally at philosophy and science (O'Neill, 1998d). However, it is not difficult to see how they might be turned against the theory of deliberative institutions. Some of the 'malpractice' of deliberative institutions on this view, in particular the existence of strategic communication within them, does not represent a falling away from norms of deliberative rationality any more than does the practice of science represent a falling away from the norms of science offered by the rationalist philosopher of science. Rather in both cases they point to the merely rhetorical force of those norms themselves.

This strong rhetorical rejection of the very possibility of rational discourse is not I believe defensible for some fairly standard reasons. There is a performative contradiction at operation in the authors' utterances in stating a strong rhetorical position. If one took seriously what they are saying in their theories, one could not take seriously their acts of saying them (O'Neill, 1995c, 1998a, ch. 12). However, while my sympathies in this dispute are entirely on the side of the deliberative theorists, a simple rejection of the rhetorical position is not adequate. While it may be true that a view which renders all deliberation an exercise in manipulation is not plausible, there are features of deliberative discourse that rhetorical studies have highlighted which point to significant weaknesses in Kantian approaches to deliberation. The strong rhetorical position needs to be kept distinct from weaker positions which allow the legitimacy of rhetorical dimensions of public deliberation without holding that all communication is strategic.[3] There may be dimensions of communication that are a proper part of public deliberation, which are consistent with respect for the autonomy of the hearer, but which Kantian models of deliberation exclude.

Two features of public discourse highlighted by rhetorical analysis are of particular importance in this regard: the role of testimony and judgements of credibility in deliberation; and the role of appeal to emotions in public discourse. Both matter from the Kantian perspective because they are potential sources of heteronomy. Testimony requires us to attend to persons, not propositions. However, to believe something on the basis of the authoritativeness of the person and not on the basis of reasons is to make oneself dependent upon another. To be moved by emotions is likewise to be governed by something independent of judgement. Moreover, that something is not ultimately the objects of our passions but other persons (Kant, 1974, 269–270). Since rhetoric is concerned both with the self-presentation of the trustworthiness of the speaker and with addressing the emotions of the hearer, it is incompatible with respect for autonomy. The source of these Kantian worries I share. However, for reasons that will become clear I also believe that the appeals to testimony and emotion are features of public deliberation that cannot and should not be eliminated. For those committed to the enlightenment values that underlie the deliberative model of democracy the question is whether these rhetorical features of deliberation are incompatible with those values. Are they incompatible with reasoned discourse and autonomy? In the remaining sections of the chapter I show that they can be rendered compatible. I do so by defending an Aristotelian account of rhetoric which rejects the Platonic contrast between reasoned discourse and rhetoric which the Kantian model inherits.

It is from Aristotle that we have the standard definition of rhetoric. Rhetoric is the art of persuasion: 'the faculty of observing in any given case the available means of persuasion' (Aristotle, 1946, Book I.2). But Aristotle also rejects the account of rhetoric as persuasion without reason. Hence he begins the *Rhetoric* with criticism of previous models of rhetoric for ignoring the place of argument in the art (Aristotle, 1946, Book I.1). Rhetoric is not set in opposition to rational argument: rather rhetoric presupposes it. Rhetoric deals in those forms of persuasion that essentially employ language. Aristotle outlines three modes of persuading through words: providing arguments themselves that are persuasive; exhibiting the authoritative and virtuous character of the speaker; moving the emotions of the audience (Aristotle, 1946, 1356a 1–20). The first mode requires knowledge of good reasoning, the second and third of ethics. Hence the claim that rhetoric is a branch of both dialectic and ethics (Aristotle, 1946, 1356a 26). All three are compatible with reason and open to reasoned appraisal. In the next section I focus on the first and second components, dialectic and character: I suggest that where the Kantian approach is unable to capture the role of trust and credibility, these features of Aristotle's position can. In the third section of the chapter I consider the role of appeals to emotion and examine the contrast between Aristotle's and Kant's accounts of reason and emotion.

10.2 Reason, authority and credibility

What is wrong with the Kantian picture of reason and authority? Consider an example. While I was writing an early draft of the arguments developed here I was listening to one of the many radio phone-ins that took place in the wake of the BSE (bovine spongiform encephalopathy) scare in the United Kingdom, and heard a standard and quite understandable worry expressed in the circumstance which have been echoed in recent debates on genetically modified organisms. A caller had just had a weekend listening to a series of scientific experts who have argued intensely with each other about the links between BSE and CJD (Creutzfeldt-Jakob disease), some saying on the one hand that there is no proof that any relation exists and some claiming on the other that there is a major health risk which is only just beginning to unfold. The caller complains: 'I just don't know who to believe. Who do I trust?' It is a standard response to a large number of problems in public policy which involve scientific expertise and are subject to controversy. Moreover, it is a problem that the various 'new' deliberative institutions such as citizens' juries and citizens' panels often address. The problem of a decline in trust in 'scientific expertise' forms the starting point of many practical applications of deliberative institutions, particularly those applied to risk. It is in part for that reason that environmental spheres have been such a prominent site for experiments in deliberative democracy.

How would the Kantian respond? Consider my caller. On the Kantian line the complaint of the caller could look like an abnegation of responsibility. Thus one response that is suggested by the Kantian model of enlightenment might be this.

> Look don't just believe when the modern scientific clerisy tell you to. That just shows a 'lack of resolution and courage to use one's own understanding without the guidance of another'. Show some maturity here. Remember the motto of the enlightenment. Have courage to use your *own* understanding.

Thus to pursue the dietary example, Kant himself claims that it is a sign of immaturity to have 'a doctor to judge my diet for me' (Kant, 1784b, p. 54). The response would be heroic, but it looks quite inadequate. It is open to the quite proper rejoinder: 'I just don't have the training to use my own understanding here. This is not something I could use my own understanding about. It is beyond my competence.' The problem here is that for all persons in most of their affairs, there exists a necessary lack of maturity which is not self-incurred. We live in society that is based upon a complex division of knowledge such that individuals depend upon the judgements of others. Evaluations of the claims made by others presuppose a specific set of social practices and education within a practice is a condition of having the capacities to engage in evaluation. For example, I do not, and without extensive training am unlikely ever to have, the capacities to make judgements about BSE or other environmental hazards. I have to take the judgements of others, for example scientifically educated friends and strangers, as authoritative. Similarly with the areas of scientific controversy of the kind that those a citizens' jury or consensus conference are often called upon to consider, most lay persons are not in a position to ascertain the truth of competing claims made to them even if they can understand them.

This is not just a special problem for the relation of scientist to citizen. It arises also within science. The person who is making judgements to which I listen, the expert, relies on the trust in others: science itself is a cooperative enterprise that involves the acceptance of the testimony of others. The biologist studying the effects of acidification of water on plants accepts already the chemist's accounts of the chemical agents involved. She does not have within her grasp all the relevant sciences: she calls on the testimony of others. The problem here is not one of time and effort, still less of courage. It is rather to do with the notions of reason and authority that runs the Kantian picture of reasoned discourse.[4] The picture of reason is a peculiarly intellectualist one, in which what is exchanged is always reason and argument directly. Judgement is supposed to be applied directly to reasons. I have to listen to the arguments of others, not to attend to their persons. In ideal conditions 'the only remaining authority is that of a good argument' (Dryzek, 1990, p. 15). However, this relies on an implausible view of the role of testimony in the justification of beliefs. Its implausibility is clear in cases of scientific controversy. In such cases citizens can rarely judge directly the adequacy of the scientific arguments offered for competing claims. Indeed, the belief that they could is a source of misunderstanding of the nature of the problem. For example, when government ministers refer to the 'evidence' offered to show that BSE is not a threat to health, that genetically modified organisms are no risk or that the best option is to dump an oil platform at sea, and talk of 'placing the facts before the public so that the public can decide for themselves', they seriously miss the point. The arguments pass me and most other citizens by. I simply wouldn't know how to appraise much of the evidence even if you gave me all the detail. I want to know not if the evidence supports this or that conclusion, but whether I have good reason to trust those who offer it. The model of the forum as a place of public argument in which the only authority is that of a good argument, in which citizens engage in the free exchange of argument, is implausible. The object of my judgement is rarely the arguments or evidence offered, but rather the credibility of those who present themselves as 'experts'. What I need is not an appeal to some unattainable 'maturity' but some criteria, 'immature' as I am in most matters, to know when it is rational to trust the authoritative judgements of others and when scepticism is appropriate. A practical problem in a democratic community is the reconciliation of necessary 'immaturity' with democratic procedures. An account of deliberative institutions has to be constructed on less intellectualist grounds. Moreover this is not just a practical problem concerning the implementation of deliberative institutions, but a problem with the picture of deliberation in 'ideal' conditions. The very concept of rational discourse as the exchange of arguments is incomplete.

Consider Aristotle's three modes of persuading through words that are constitutive of rhetoric: providing arguments themselves that are persuasive; exhibiting the authoritative and virtuous character of the speaker; moving the emotions of the audience (Aristotle, 1946, 1356a 1–20). While the first, the presentation of argument, belongs to dialectic, it is not distinct from the second. Aristotle famously distinguishes logical and dialectical argument. Logic deals in demonstrative inferences from premises that are 'true and primary' that is, 'believed not on the strength of anything else but themselves' such that 'it is improper to ask any further why and wherefore of them' (Aristotle, 1928, Book I, ch. 1). In certain domains, notably mathematics, demonstrative inference is to be expected. However, other domains demand different levels of rigour: 'the educated person seeks exactness in each area to the extent that the nature of the subject allows; for apparently it is just as mistaken to demand demonstrations from a rhetorician as to accept [merely] persuasive arguments from a mathematician' (Aristotle, 1985, Book I, ch. 3). Dialectical argument employs persuasive arguments, or to use Burnyeat's apt phrase 'relaxed arguments' (Burnyeat, 1994), that is arguments that may not meet the norms of full deductive validity, but make justifiable claims on the rational mind. It reasons from endoxa, authoritative opinions (Aristotle, 1928, Book I, ch. 1). Such opinions form the starting point for inquiry: we need to begin from 'what is known to *us*', not from propositions known 'unconditionally' which form the starting point for logical inquiry (Aristotle, 1985, Book I, ch. 4).

The feature of dialectical argument that is of particular note here is that it takes authoritative utterances as its starting point. It is the art of interrogating the conflict between propositions thus offered to us by others. That this is our starting place points to the significance of the second component in Aristotle's account of rhetoric – the presentation of character. The analogy that Aristotle presents in the *Metaphysics* (Aristotle, 1908, 3.1) between a person resolving the conflict between the variety of opinions and theories presented to him and the judge deciding in a lawsuit can be understood as a strong one. In both the character of the person to whom one is listening matters:

> the orator must not only try to make the argument of his speech demonstrative and worthy of belief; he must also make his own character look right and put his hearers, who are to decide, in the right frame of mind.
>
> (Aristotle, 1946, II.1)

In making a judgement we need to judge the credibility of the speaker. That credibility has both epistemological dimensions – the speaker must have good sense and be reliable in the formation of judgements – and ethical dimensions – the speaker must have the moral character that allows us to trust their utterance and there must be grounds for believing that the speaker is not inclined to impart falsehoods to his audience (Aristotle, 1946, II.1).

This emphasis in Aristotle on the role of a character in the appraisal of knowledge claims provides a more plausible picture of the nature of deliberative institutions than that which has come to us from the Kantian tradition. Evaluation of the character of sources of belief is central to our most basic knowledge claims. Most of what we believe is founded on testimony of others. Since we cannot confirm their claims themselves, we have to rely on judgements about the credibility and reliability of sources. As noted earlier in the chapter, such judgements are unavoidable. The division of knowledge in society of itself entails that we have to trust the testimony of others: the fact that most knowledge claims can be evaluated only by those with the requisite training within some discipline entails that most will remain opaque to our best judgements. Hence it is the credibility of the sources of knowledge claims that is the object of appraisal not those claims directly. It provides the central problem of practical epistemology in the democratic forum. The citizen needs to rely on the authority of others to make judgements. The problem she faces is knowing how and where to apply proper scepticism to testimony (O'Neill, 1993a, ch. 8).

In making this point the account of rhetoric also points to what remains valuable in recent sociology of scientific knowledge once the relativism and global anti-realism often associated with it are rejected. Much that goes by the name of anti-realism and relativism in the social studies of science is in fact aimed at a form of intellectualism which makes the credibility of a claim purely a matter of how well it is supported by other propositions. The object of criticism is that outlined by Shapin (1995):

> The credibility and the validity of a proposition ought to be one and the same. Truth shines by its own lights . . . Once upon a time . . . students of sciences . . . believed that truth was its own recommendation, or if not that, something very like it. If one wanted to know, and one rarely did, why it was that true propositions were credible, one was referred back to their truth, to the evidence for them, or to those methodical procedures the unambiguous following of which testified to the truth of the product.
>
> (Shapin, 1995, pp. 255–256)[5]

Once it is admitted that the issue is one of the credibility of knowledge-claims and not their truth, then it is quite proper to point to the variety of social causes of credibility or even their social construction. It does not follow that truth or knowledge are thereby social constructions unless one already accepts a naive identification of credibility and truth.[6] Neither does it follow that credibility is a sufficient condition of knowledge nor, without some additional argument, that it is a necessary condition – that 'no credibility, no knowledge' (Shapin, 1995, p. 257).

In the modern world the issue of credible testimony has an institutional and political focus. We live in a world in which testimony is offered to us by strangers, by 'spokespersons' and 'experts', who call upon us to believe what they say on the basis of certification from institutions – academic, industrial, commercial and political.[7] The different institutional dimensions of evaluation are of significance here. What institutions deserve epistemological trust in what conditions? What is required as an answer is not a series of decision rules but a political epistemology concerning conditions of trust, and a corresponding social and political theory about its institutional preconditions. The association of evaluative practices with positions of social power and wealth for example induces quite proper scepticism about their reliability. Indeed, there is I believe a good epistemological argument for social and political equality founded upon conditions of trust (O'Neill, 1993a, ch. 8). In the absence of ideal conditions, we require tools to scrutinise institutions, tools that will include an everyday sociology of science.

Much public debate in current conditions is in fact at the level of how to assess credibility. Thus to return to my confused caller worried about who to trust on BSE, one reply he received on the phone-in was eminently sensible: Don't trust those who have an axe to grind, those who are tied to the food industry and attempt to reassure you. Seek those who are independent. And remember there is a difference between not proving there is a connection between BSE and CJD and proving there is no connection. The arguments make no mention of evidence or specific scientific arguments for or against the connection. They give good rules of thumb about who to trust and how to interrogate arguments with some general principles of good reasoning. It calls upon an understanding of the conditions in which trust in the authoritative utterances of others is rational and tools of proper scepticism. They are the tools of scepticism that are employed in citizens' juries where they interrogate the testimony of experts.

To defend this rhetorical model of the forum is not to advocate any irrationalism or authoritarianism, nor to reject the values of the enlightenment, in particular that of autonomy. It is to reject a particular conception of reasoned deliberation and epistemological maturity which denies that there is ever an occasion for accepting claims on testimony or authority (O'Neill, 1998a, ch. 7). In its place one has a less heady picture of maturity that is more akin to the skills an agent in a public forum requires. She is able to reason well for herself, but knows when her own reason is insufficient; she is not credulous nor willing to accept all and any propositions put to her by putative authorities; she is able to judge whose testimony is reliable, whose is not; she knows when and how to be suspicious; she is versed in the practical art of suspicion. The skills of the mature autonomous agent are not, on this account, merely those of the good logician – although these are necessary. They are those of the person who knows when and where it is reasonable to trust claims that call on authority.

10.3 Autonomy, maturity and emotion

I have thus far focused upon the role of dialectic and character in rhetoric. I have left aside its other component – the appeal to the emotion. It might be argued here that this does render rhetoric incompatible with reason and autonomy: the appeal to emotion overrides the call upon reason. Moreover it might be argued that this is central to the aim of rhetoric, for what classically distinguishes rhetoric from dialectic is that it aims to move an audience to action. Dialectic has theoretical aims, rhetoric practical. The difference between the two arts maps onto the traditional distinctions between being instructed and being moved (Augustine, 1952, IV, ch. 13, para. 29). Kant's criticism of rhetoric inherits the view that rhetoric moves us by appeal to the passions and not judgement, where these are taken to be independent of each other. On the Kantian view emotions are feelings, capacities for psychological sensations of pleasure or pain. These might sometimes accompany cognitive states but they are themselves without any cognitive dimension

(Kant, 1964, 210–211).[8] Since the passions are non-cognitive it follows that reasoned discourse cannot appeal to the emotions. Hence it follows that appeals to the emotion move us without engaging our judgements. As such they render the agent passive, someone who is impelled to act without rational deliberation and choice. Emotions are psychological states that move us, and sometimes overwhelm us. They are not states that we freely choose. Hence, to move an agent by appeal to emotions is to render them unfree. Thus rhetoric is rejected as inconsistent with reason and autonomy.

The view that the rhetorician's appeal to emotion moves them to act without reasoned judgement is an ancient one to be found in both classical proponents and critics of rhetoric. One of Aristotle's most important contributions to the study of rhetoric lies in his criticism of the account of the emotions that underpins this view. While Aristotle opens the *Rhetoric* with the complaint that previous treatises focus only on tricks that sway the emotions without the use of proper judgement, he is not thereby committed to the anti-rhetorical position that any rhetorical appeal to the emotions is simply an exercise in the non-rational swaying of the mind. It is not so because the emotions themselves are not non-cognitive states, but are constituted and individuated by beliefs. As such they answer to reason and are open to rational persuasion. This point is a central part of what distinguishes the Aristotelian defence of rhetoric from the anti-rhetorical tradition in philosophy.

Aristotle's theory of rhetoric presupposes his theory of the emotions. Indeed the *Rhetoric* is the main source for this theory. For Aristotle, the emotions are open to the appraisal of reason. Such appraisal of emotion is possible because the emotions are partially constituted by beliefs. Thus consider Aristotle's definitions of anger and pity:

> Anger may be defined as an impulse accompanied by pain, to the conspicuous revenge for a conspicuous slight directed without justification towards what concerns oneself or towards what concerns one's friends . . . Pity may be defined as a feeling of pain caused by the sight of some evil, destructive or painful, which befalls one who does not deserve it.
>
> (Aristotle, 1946, Book II, ch. 2, 1378a 31–33; ch. 8, 1885b 12–15)

Whatever the adequacy of the specific definitions 'anger' or 'pity' in the modern sense, they are right in highlighting the way that beliefs are constitutive of the emotions. Part of what it is for an emotion to be that of pity is that it is directed towards a person of whom it is believed harm has fallen. Because emotions have beliefs constitutive of them, they are open to appraisal. They can be appropriate or inappropriate, felt at the right time, of the right things, for the right reasons or not (Aristotle, 1985, Book II, ch. 6, 1106b 18–23). I feel anger because a distant relative forgets my birthday – the emotion is irrational, since I know that no slight is intended. I feel anger at the dismissal of a colleague: the anger is quite rational for the harm is intentional and unjustified.

More significant in the present context, it follows that the emotions are not deaf to reason, but open to the rational persuasion (Aristotle, 1985, Book I, ch. 13).

Emotions are constituted by beliefs and can be roused by addressing these. I rouse your anger by pointing to unjustified harm the logging company does. I placate your anger at farmers who destroy a valuable habitat by placing their action in the context of grinding poverty: your anger is redirected to those responsible for this, and in its place you are moved to pity the farmers and their families. Such appeals are the stuff of everyday discourse and public debate. They are also central to the art of rhetoric. On this view rhetoric is not simply an attempt to sway the emotions in an irrational or arational way. It rather involves giving grounds for belief, and can be appraised in virtue of doing so. The putative incompatibility of rhetoric with the public use of reason is undermined.

A further point to note here is that given this view of the emotions, the role of the emotions in motivating action is not simply a case of it supplying a non-rational drive or impulse to movement. Rather emotions involve perceptions and judgements about what matters in particular cases. To feel pity for the farmers involves the perception of their suffering, the judgement that this is undeserved, and pain felt in response to that suffering. In consequence we are moved to act against those responsible for their condition. We are not moved in the sense of being hit by some non-rational impulse to act. Correspondingly, to educate the emotions is to develop cognitive capacities of perception and judgement, not simply behavioural tendencies to movement.

This appeal to the rationality of emotion also points to a wider model of maturity than that in the Kantian picture. It would be false to say that Kant's picture of the 'mature' agent lacks any emotional dimension. The central virtue of the mature individual is courage – the mature agent is one who dares to use his own understanding. Courage properly specified has an emotional component: it involves remaining firm in the face of what is fearful. Moreover, for Kant, while emotions are non-rational, the cultivation of emotions is part of the development of moral maturity (Kant, 1948, 393–394).[9] However, since the emotions are non-rational, their role is subsidiary. Maturity is still primarily a matter of intellectual judgement, the capacity to use one's own understanding. On the Aristotelian model the capacity to have the right emotions on the right occasions is a central mark of moral maturity.

Having thus far praised Aristotle's response to the problem of reason and the emotions it has also to be admitted that the position is not entirely satisfactory. There is something as heroic and unrealistic about the Aristotelian version of maturity as there is about the Kantian. The picture of a perfect harmony between practical wisdom and emotional response is implausible of empirical agents. Agents who are always moved appropriately, who feel emotions 'at the right times, about the right things, towards the right people, for the right end, and in the right way', are not agents we are ever likely to meet. The kind of universal capacity for appropriate responses may simply not be the kind of capacity that humans can exhibit (O'Neill, 1997b). Moreover, as Kant notes, there is also a passivity about emotions. On occasions we are overwhelmed by emotion, and moved to act in ways in which are against our better judgements. The latter point clearly to matters for public deliberation. It forms a defensible core to the Stoics' objection to Aristotle's account of the emotions which Kant in part inherits. The Stoic argues that emotions are not the kind of states that are always open to moderation. Emotions run away with us and move us to excess. It is an objection that Seneca presses in pointing to the disastrous role the appeal to emotions like anger can have in public life (Seneca, 1995; see also Nussbaum, 1994, ch. 11).[10] The demagogues of the ancient world had no worthier record than do those of the modern world. The objection has clear power. However, the Stoic response does not. Neither the Stoic view of the emotions – that they are false judgements – nor the Stoic cure to such dangers – the elimination of the emotions – are plausible. Indeed they fall together. The emotions do not always involve false judgements about what is of value – pity or anger felt appropriately involve true judgement and perception, for example of the conditions of our farmers and their plight. Because emotions are capacities for proper judgement and concern about what matters in private and public life, they are not or should not be eliminable from public life. The proper response to the Stoic objections is that we have to accept vulnerability of public institutions to the sometimes damaging consequences that appeals to the emotions can bring and design institutions to guard against these. It is not to attempt to eliminate the emotions. The same response is owed to the Kantian proponents of deliberative democracy. The emotions are open to reasoned dialogue, and they move us through judgements. The response to the dangers of rhetorical appeals to the emotions that impel us to false judgement is not to imagine that good deliberation could do without appeal to emotions. Rather it is to design institutions to minimise the dangers that follow.

Recent deliberative models of democracy are founded in enlightenment values: they offer a picture of democracy as a conversational process which forms mature citizens who are able to judge for themselves. The great virtue of the Kantian tradition in political thought has been its articulation of an account of deliberation as part of the educative process of enlightenment. It needs to be defended against the recent criticisms of public choice and rhetorical critics who aim to show that all speech is strategic. However, in responding to those rhetorical critics, it has to be acknowledged that the intellectualist account of deliberation it offers is inadequate. Political deliberation cannot be adequately understood in purely intellectualist terms as the exchange of arguments. Aristotle's rhetorical account of public deliberation is the more plausible: judgement is applied to the credibility and trustworthiness of persons and institutions; speech addresses the emotions. However, this need not entail the abandonment of the enlightenment project. The virtue of the Aristotelian position is that it allows us to develop an account of deliberative institutions as rhetorical without giving up the core ideals of the enlightenment. Reason, maturity and rhetoric are compatible.[11]

11 Representing people, representing nature, representing the world

11.1 Representation and the environment

At the heart of a series of debates about the environment are problems of representation. At the general level there are problems about those who are without adequate representation in political and economic choices about environments in which they will live. Some are unrepresented or under-represented for contingent reasons – given a different political and economic order they could be better represented: the working class, peasants, many groups of women, marginalised indigenous groups, and ethnic minorities. For others there are problems about the very possibility of representation: future generations and non-human beings pose particular problems of representing those who cannot speak and have in that sense no possibility of voice or presence in processes of environmental decision making. The absence of such representation raises major problems concerning the ethical and political legitimacy of decisions made in the absence of their voice. At the same time it raises problems also for forms of environmental action and advocacy that are legitimised by appeal to the claim that protagonists are speaking and acting on behalf of those who are without voice.

Alongside these general issues there are more particular problems of representation posed by different institutional structures and arrangements employed in environmental decision making. Central to the case against the use of market and surrogate market prices as a basis of environmental choice are failures of representation. They leave the poor under-represented: since willingness to pay is income dependent the use of raw willingness to pay measures will give greater weight to the preferences of the well-off – 'the poor sell cheap' (Guha and Martinez-Alier, 1997). The interests of non-humans and future generations cannot be directly represented at all through willingness to pay and they are indirectly represented at best precariously through the preferences of current consumers (O'Neill, 1993a, ch. 4). As I noted in Chapter 9, deliberative institutions are often presented as a response to these failings. Hence the recent revival of deliberative democracy and the development of a variety of 'new' formal deliberative institutions such as citizens' juries, citizens' panels, focus groups, in-depth discussion groups, consensus conferences and round tables. Such deliberative institutions are claimed to resolve some of the problems of inadequate representation involved in surrogate market methods. The distribution of resources and property rights is not presupposed as it is in market methods, but is itself an object of public deliberation. The publicness condition on deliberation, that reasons must be able to survive being made public, is taken to force participants to offer reasons that can withstand public justification and hence to appeal to general rather particular private interests.

However, the deliberative alternatives have their own problems of representation. Willingness and capacity to say and to be heard is unevenly distributed across class, gender and ethnicity. The problem of future generations and non-humans, whose direct voice is necessarily absent, is a recognised source of difficulty for deliberative theories of democracy (Dryzek, 1992; Goodin, 1996; Jacobs, 1997; Eckersley, 1999). However, it is not just with non-humans and future generations that the problems of representative legitimacy arise for deliberative theories. In recent experiments in deliberative institutions – citizens' juries, in-depth discussion groups, consensus conferences and the like – the institutions are normally small and are open to the criticism that they are 'unrepresentative' of the populations to be affected by decisions (Kenyon et al., 2001). Moreover, this problem may be thought to be a necessary feature of deliberative institutions: the quality of deliberation improves in smaller fora, the quality of representativeness in larger fora and the two demands pull in the opposite direction.

11.2 The Alejandro solution

What makes for good representation – or at least adequate representation – for deliberative institutions? What institutional forms will suffice in what conditions? I want to start to answer those questions with a passage from a story by Borges, 'The Congress':

> Don Alejandro conceived the idea of calling together a Congress of the World that would represent all men of all nations . . . Twirl, who had a farseeing mind, remarked that the Congress involved a problem of a philosophical nature. Planning an assembly to represent all men was like fixing the exact number of platonic types – a puzzle that had taxed the imagination of thinkers for centuries. Twirl suggested that, without going farther afield, don Alejandro Glencoe might represent not only cattlemen but also Uruguayans, and also humanity's great forerunners, and also men with red beards, and also those who are seated in armchairs. Nora Erfjord was Norwegian. Would she represent secretaries, Norwegian womanhood, or – more obviously – all beautiful women? Would a single engineer be enough to represent all engineers – including those of New Zealand?
>
> (Borges, 1979, pp. 20–22)

The story ends with an echo of another of Borges' stories on development of cartography which attains its highest point when the perfect map is identical to the area it maps: 'the College of Cartographers evolved a Map of the Empire that was of the same Scale as the Empire and that coincided with it point for point'.[1] Similarly the only adequate congress of the world is discovered to be the world itself.

> 'It has taken me four years to understand what I am about to say' don Alejandro began. 'My friends, the undertaking we have set ourselves is so vast that it embraces – I now see – the whole world. Our Congress cannot be a group of charlatans deafening each other in the sheds of an out-of-the-way ranch. The Congress of the World began with the first moment of the world and it will go on when we are dust. There's no place on earth where it does not exist'.
>
> (Borges, 1979, p. 32)

The characters go back out into the city 'drunk with victory', to take part in the life of the Congress in which all people and all things are represented perfectly by themselves.

It would be nice to think that don Alejandro had the last word on the subject, and in the design of institutions for representation the solution should be that we should simply go for a walk outside. Indeed one version of the idea represents a form of democratic anarchism that is attractive. However, I wouldn't propose to use Borges' character Alejandro to design representative institutions anymore than I would use his cartographers to design maps. Maps are better when not identical to their original.[2] The Congress of the World is not best represented by the world itself. In both cases, the solution rather misses the point of representation and the work it does in our lives.[3] Borges' discussion remains however important. First, there is often a background assumption in discussions of representation that the perfect representation would be the person or object itself. Critics of 'essentialism' sometimes complain that representatives do not map all the characteristics of those they represent. They could not and this is not always a problem. Every representative institution is unrepresentative over some dimensions. No representation captures everything about those represented. It couldn't and it is not always a problem. I don't complain that my Ordnance Survey map doesn't mark every stone. It doesn't matter that men with red beards do not have a spokesperson in the United Kingdom. Indeed, given the obvious candidates, I'm rather glad they don't. I do complain when my walking maps are indifferent to scale. There are proper complaints that large groups of the population lack any adequate representation in political life, for example low-paid and retired workers. It is a criticism of a social scientific survey that it is unrepresentative of the class or ethnic composition of the group it studies. Second, as we shall see later, Alejandro solutions threaten some approaches to the political problem of representation. For example, certain versions of the politics of difference – those which combine the need for the presence of all identities in political decisions with the denial of 'essentialism' and with it of the claim that anyone can stand in for anyone else – very quickly tend to Alejandro solutions.

What then are the criteria for saying who or what should be represented and whether representation is adequate? Here again, rather reluctantly, I want to depart from one suggestion from the Borges story with which I opened. The problem is not one of a philosophical nature, at least in the sense that Twirl suggests, that of correctly fixing on Platonic forms – the reason why being 'a man with a red beard who sits in armchairs' wouldn't normally be a relevant characteristic is not because it wouldn't appear in a theory of forms. What then is the reason? There are at least two independent answers that might be made to that question. The first concerns prediction and explanation. The category of men with beards who sit in armchairs is not one that would appear in any significant generalisation in the social sciences. There are no significant statistical correlations one might discover with other social and psychological variables – beliefs, attitudes, behaviour and so on – and correspondingly no predictions one might make about them on the basis of that description. Neither is it the case that those features of the person do any explanatory work in any explanation of beliefs, attitudes, behaviour and so on. Second, there is a normative answer, that under that description there are no interests, judgements, or concerns that demand representation. The two answers are logically independent, and which we are considering matters to the issue of adequacy of representation.

11.3 Representation: social scientific or political

Consider the objection that is sometimes made about of many recent experiments in deliberative institutions such as citizens' juries, citizens' panels, focus groups and in-depth discussion groups, that they are 'unrepresentative' of the larger population (Kenyon et. al, 2001). What is the objection? One version of the objection is predictive and explanatory, and brings the discussion of deliberative institutions within the ambit of a larger discussion of the role of small-scale qualitative research in the social sciences. The objection runs that in small deliberative processes the samples are too small to be statistically representative and hence one could make no significant generalisations about behaviour of larger populations on the basis of the populations employed.[4] For that reason some social scientists, especially among economists and psychologists for whom such statistical questions are central, tend to be more sceptical of their use. The economist in doing contingent valuation will ask questions about the relationships between willingness to pay and other social variables, such as class, gender and ethical beliefs. The social psychologist might ask whether different measures of ethical belief, e.g. 'anthropocentric' and 'biocentric' cluster in a way that will allow generalisations about behaviour. The categories that are significant on this view tend to fall out of statistical analysis. When it comes to small groups typical of deliberative institutions, no such analysis is possible. The proponent of this view can then make certain standard claims about the usefulness of their approach to policy makers which deliberative models cannot realise. The policy maker needs to know how a population will respond to various policy instruments if they are to be effective and the social scientist provides the generalisations to support this. In contrast, the argument runs, small-scale qualitative techniques in social science, if they are supposed to tell us something about how citizens are going to respond, will be inadequate. They provide no basis for generalisations.

The argument appeals here to a particular view of what social sciences should be doing and what its role in policy making should be. It appeals to something like a deductive nomological model of social science. The aim of social science is to discover significant law-like regularities in a population: hence the variety of statistical techniques to ascertain whether significant correlations exist. Explanation of a particular phenomenon consists in bringing a particular state of affairs under a law-like generalisation. The role of the expert is to offer the policy maker advice on the effectiveness of different policies, given some policy aim. For example, given that the aim is maximisation of preference satisfaction, then, the argument runs, cost-benefit analysis will give the policy maker precise information about which of a set of policies does so effectively. There may be other aims in the policy arena that this has to be subsequently squared with, but that is a job for the citizen and politician not the social scientist. The criterion of effectiveness is what guides the role of the scientist in the policy-making arena.

This line of argument is open to some fairly standard responses – that it involves a mistaken view of social science and an unacceptably technocratic view of politics. The objections are well developed elsewhere, but it is worth briefly outlining their main contours.[5] As a view of social science the deductive nomological model fails to offer an adequate account of either interpretation or explanation in the social sciences. Interpretative activity is already implicitly presupposed in arriving at statistical generalisations between social variables.[6] A number of criticisms of contingent valuation of environmental goods discussed in Chapter 1 have their basis in the failure of valuation techniques to take this interpretative dimension seriously. In neoclassical underpinnings of contingent valuation willingness to pay is treated as a neutral 'measuring rod' of the utility a person expects to receive from an object.[7] However, that characterisation is too thin. Acts of buying and selling are not like exercises in the use of a tape-measure. They are social acts with a particular social meaning and many protest bids to willingness to pay questions, such as refusals or zero bids, are a consequence of those social meanings. Particular social relations and ethical commitments are expressed through a refusal of market exchange – to accept a price on certain goods is an act of betrayal (O'Neill, 1993a, ch. 7; 1998a, ch. 9). The deductive nomological model also fails to offer an adequate account explanation. Explanation does not consist in just bringing particular events under law-like generalisations. It involved the establishment of underlying causal powers and processes which are necessary to distinguish between accidental and causal regularities. The existence of statistical correlations does not as such establish the causal story required for adequate explanation.[8]

Correspondingly, the significance of qualitative data in social sciences of the kind small deliberative institutions offer is that it potentially offers both interpretative and explanatory depth that large-scale statistical studies cannot provide. It offers interpretative depth, providing insights into the meaning of the responses offered. In aiming at hermeneutic depth, one has to treat the subjects of interpretation as potential partners in dialogue, as communicative agents, not objects (Habermas, 1984). It also offers possible explanatory depth, of insights into why individuals and groups respond as they do. Inference from particular qualitative case studies to general claims is licensed not by statistical regularity between variables, but rather by appeal to bridging theoretical explanatory or interpretative claims (Mitchell, 1983). None of this is to rule out a role for the search for statistical regularities in the social sciences. It serves rather to point to a proper but more limited role in social scientific research than the deductive nomological account supposes.

These interpretative and explanatory aims that underlie qualitative research are still compatible with the view that the role social science in the policy making is primarily about effectiveness in the delivery of policy – of giving the policy maker better intelligence as to how subjects will respond to different policy signals. Indeed, many qualitative research techniques are used for clearly strategic purposes in which the primary background value is effectiveness. Consider for example focus groups employed by both market researchers and political parties. However, the shift to deliberative institutions in research is often associated with a shift in focus in the values that inform its use – from effectiveness to democratic legitimacy. Norms of communicative rationality rather than instrumental rationality guide the process.[9] On this view social science involves a critical engagement with an audience that extends democratic dialogue. It is a critical engagement in that interpretations and explanations offered in dialogue can conflict with those initially held by those with whom it engages – it involves not just recording given views and attitudes, but ideally their transformation through conversation. It extends democratic dialogue by including in deliberation voices that would otherwise be unheard. Part of the promise of some forms of qualitative research is that it includes directly the words of those who would otherwise be unheard. Thus, in outline, goes the story. And there is much in this response with which I have some sympathy. However, it doesn't as such resolve the problem of representativeness. It rather transforms the nature of the problem: the adequacy of representation becomes a normative problem and not just an empirical and explanatory problem.

11.4 Political theory and the problem of representation

The normative problems concerning the legitimacy of democratic representation are not identical to the empirical and explanatory problems raised by statistical representation. Those normative questions have their own long history in political theory (Pitkin, 1967; Birch, 1972). They typically include the following questions.

Who is being represented in the political or decision-making process? Under what descriptions are they being represented? Is it individuals as individuals, as members of particular functional groups or economic classes, as bearers of particular identities, cultural, ethnic, or gender?[10] Is representation limited to current generations of humans or should it be extended to include representation of future generations or nonhumans? The answer to any such question has to appeal to normative criteria. It is a matter of who should count under what description and for what ends, not a question of simple statistical significance. Moreover, it is worth noting, there are also descriptions under which it might be objectionable to be represented in virtue the attitudes they embody – consider for example, the categories of 'Norwegian womanhood' or 'beautiful women' in the passage from

Borges. Other categories are ethically irrelevant e.g. 'men with red beards who sit in armchairs'.

What is being represented? Is it the particular interests of different individuals or groups? Is it their common interests? Does it include their values, their opinions, their preferences, their will, or their identities?

Who is doing the representation? What relation do representatives have to the represented? Should it be the case, as it is within some traditions of egalitarian thought, that the representatives of some class or group should share a common identity with those they represent?

What is the source of the legitimacy of the representation? Three important answers to that question have particular relevance here.

Authorisation and democratic accountability

The claim that authorisation is central to representation is given its starkest formulation in Hobbes who makes it the whole of representation (Hobbes, 1968, ch. 16). Representation is modelled on the legal attorney who speaks or acts on behalf of a client on being authorised to do so. However, in a number of democratic theories of representation authorisation is embodied in democratic election (Plamenatz, 1938; Pitkin, 1967, pp. 42ff.). As such, authorisation is tied to the accountability of the representative to the represented, a feature absent in Hobbes' account of authorisation. Thus on one liberal model of representation, it is the interests of individuals that are represented through the act of authorisation embodied in the vote. More radically within socialist and egalitarian politics authorisation is often associated with the representation of class interests through recallable delegates. Authorisation is itself agnostic on the question of who does the representing: the representative when authorised can be entirely different in characteristics from the person represented. A lawyer can represent children without being a child. And in certain contexts one may prefer that X speaks for you, because of features you do not share – for example, X is more articulate as a shop steward in an industrial tribunal.

Presence/shared identity

A feature of much feminist and socialist theory is that in the context of political representation a pure authorisation model is rejected. Who does the representing matters. Thus in 1789 the following by women claiming a place in the Estates General:

> Just as a nobleman cannot represent a plebeian and the latter cannot represent a nobleman, so a man, no matter how honest he may be, cannot represent a

> woman. Between the representatives and the represented there must be an absolute identity of interests.
>
> (cited in Phillips, 1997, p. 175)

The same thought underlay the principle in the socialist movement that the emancipation of the working class must be their own work, and correspondingly the rejection of representation from any other class. The thought that particular groups demand representation by those who share a common identity has become central to what Phillips calls 'the politics of presence' (Philips, 1995, 1997; cf. Kymlicka, 1996; Gould, 1996). It underpins, for example, the demand for quota systems in modern electoral systems. The claim that shared identity is a necessary condition of adequate representation doesn't of course preclude the necessity of authorisation or accountability. That someone shares an identity with me under some description does not entail they can legitimately represent me in the absence of my authorisation.[11]

Epistemic values

A third source of appeal to legitimacy is that of knowledge, expertise or judgement that allows an individual to speak or act on behalf of some group. That argument is often developed in ways that are in tension with a representation legitimised through shared identity. Consider for example the view that there are certain individuals who through knowledge have a better grasp of the objective interests or good of some group than others in that group and that this knowledge legitimises their representative status. Versions of that appeal are to be found in theorists as different as Burke, Mill,[12] and Lenin.[13] However, knowledge and presence need not always be in tension. One reason why a shared identity might be held to matter in representation is that similar experiences are a condition of proper knowledge of the interests and aspirations of that group.

11.5 Deliberative democracy and the sources of legitimacy

How do deliberative institutions fit with different answers to these questions about representation and its legitimacy? New deliberative institutions can be understood in part as experiments in deliberative democracy. In its minimal sense deliberative democracy refers to the view that democracy should be understood as a forum through which judgements and preferences are formed and altered through reasoned dialogue against the picture of democracy as a procedure for aggregating and effectively meeting the given preferences of individuals. As such the deliberative theory does rule out certain answers to the question of *what* is being represented. It is not simply given preferences that are represented in debate.[14] Preferences are rather formed through debate. Deliberation in this context also needs to be distinguished from processes of negotiation or bargaining between different particular private interests, important as such processes are (O'Neill, 1997a, pp. 84–85; Elster, 1998b). What is being represented has to be understood as a judgement that is open to reasoned debate. Deliberative democracy has then implications then as to what is being represented.

However, when it comes to the other questions I raised above, deliberative institutions are consistent with several different answers. In particular, deliberative democracy is compatible with different answers to the question of the sources of legitimacy. It is, for example, quite possible to understand Edmund Burke's famous address to the electors of Bristol as offering a deliberative theory of democracy, underpinned by epistemic claims to legitimacy:

> Your representative owes you, not his industry only, but his judgement; and he betrays, instead of serving you, if he sacrifices it to your opinion . . . [G]overnment and legislation are matters of reason and judgement, and not of inclination; and what sort of reason is that, in which the determination precedes the discussion; in which one set of men deliberate and another decide . . . Parliament is not a *congress* of ambassadors from different and hostile interests; which interests each must maintain, as an agent and advocate, against other agents and advocates; but parliament is a *deliberative* assembly of *one* nation, with *one* interest, that of the whole.
>
> (Burke, 1774, pp. 95–96, emphases in the original)

This is a deliberative model of political institutions in the sense that it sees political processes not as the aggregation of preferences or inclinations, nor as a process of negotiation between different private interests, but as a matter of arriving at common judgements on common interests founded on reasons and argument. However, it is clearly not the model of deliberative institutions that informs the recent theory and practice of deliberative democracy. The difference from this older tradition is sometimes marked by the attempts to marry deliberative theory with 'inclusionary' models of democracy.

The appeal to an ideal of inclusion links the arguments about representation to some of the themes in the politics of presence – that deliberative institutions should be such that they give equal access to all relevant voices by including directly representatives of different relevant identities. Wider inclusion is defended as a source of proper deliberation in which the widest range of different relevant views are heard. Typical is Sunstein (1997):

> [D]eliberative processes will be improved, not undermined, if mechanisms are instituted to ensure multiple groups have access to the process and are actually present when decisions are made. Proportional or group representation . . . would ensure that diverse views are expressed on an ongoing basis in the representative process, where they might otherwise be excluded.
>
> (Sunstein, 1997, p. 169)

The argument runs that a deliberative justification of the presence of different identities offers a more powerful defence than that based on a model of bargaining: 'the primary process of access is not to allow each group to get 'its piece of the action' – though this is not entirely irrelevant – but instead to ensure that the process of deliberation is not distorted by a mistaken view of a common set of interests' (Sunstein, 1997, p. 169). It is through confronting a range of arguments and views that preferences and judgements change. Inclusion then is justified by the nature of deliberation itself. The presence of different identities may be considered part of the ideal conditions in which consensus under the sole authority of the better argument emerges.

However, the link between experiments in deliberative democracy and the politics of presence raises a number of problems of legitimacy which echo Twirl's questions in the Borges story with which I opened. Individuals have a complex set of identities under different descriptions. The issues, as to which of these count, under what categories different groups are included and how legitimately particular individuals can be said to stand for or speak on behalf of others becomes increasingly problematic:

> What . . . is an appropriate mechanism for dealing with political exclusion? Can Asians be represented by Afro-Caribbeans, Hindus by Muslims, black women by black men? Or do these groups have nothing more in common than their joint experience of being excluded from power?
>
> (Phillips, 1997, p. 181)

Or to take examples from experiments in deliberation, in focus groups comprising Blackburn Asian women, or unemployed young men from Morecambe, who are they supposed to represent (Macnaghten et al., 1995)? Of whom should a retired worker on a citizens' jury about the re-creation of a wet fen site be taken to be representative (Aldred and Jacobs 1997)? The very descriptions used in these contexts by the social analyst force a representative status on participants which may be contested, indeed by those individuals themselves. The contestation is proper. Simply being a member of a particular group does not entail that one is a representative of that group. [15] And if these quite proper questions are combined with what is often called a radical 'anti-essentialism', which denies the very idea that there are any characteristics that are shared by some group, then the subsequent 'distrust [in] the notion that anyone can "stand in" for anyone else' (Phillips, 1997, p. 180) may very quickly lead to an Alejandro solution in which representation by others is impossible. This in part points to problems in anti-essentialism, but I leave those aside here (O'Neill, 1995d, 1998a, ch. 1; see also Cole, 1920). The problems of legitimacy of representation in recent deliberative institutions remains even if radical anti-essentialism is denied.

The problems are made more acute where authorisation and accountability as sources of legitimation are absent. There is a clear sense in which in many recent experiments in inclusionary deliberative institutions such as citizens' juries, citizens' panels, consensus conferences, in-depth discussion groups or focus groups, none of the participants are authorised to speak for any group they are taken to 'represent'. Nor are they accountable to those groups. Hence in what sense can they be taken to be a legitimate representative? Indeed in many cases their status of being representative of some group is not chosen by the participants themselves in deliberation. For that reason, it is not clear that they can be said to have the same sources of legitimacy that are appealed to in traditional democratic representative institutions.

There may, however, be other good reasons for taking at least some forms of the new deliberative institutions to be part of a democratic social and political order. One response is to insist upon a weaker role for such institutions within the democratic process, say to the formulation of options and possible recommendations, allowing for other forms of accountability to be retained in the decision-making process (Jacobs, 1997, p. 224). Another more promising argument in favour of a larger role for some forms of deliberative institution is to appeal to the Athenian practice of having political positions filled by lot as part of the democratic order, a position that Burnheim (1985) has revived under the name of 'demarchy'. The justification of the procedure lies in the idea of democracy as a social order of equals where citizens take turns in positions of power, in which, as Aristotle puts it, 'they rule and are ruled in turn' (Aristotle, 1948, II.ii). Among the main arguments in favour of the position is that, unlike most other institutional arrangements, power is not distributed to those who desire it and both power and responsibility circulate among citizens. In addition the approach offers a counterweight to the power of experts in decision making.[16] However, such Athenian solutions still require some institutional arrangements for the accountability of deliberative institutions to a wider public. Burnheim suggests otherwise: 'Demarchic leaders would not be accountable because they would not be eligible for reappointment' (Burnheim, 1985, p. 167). However, the reasoning here is unconvincing. The argument could and should run in the other direction; that is, because demarchic leaders are closed to reappointment, other forms of accountability need to be developed.

11.6 Giving voice to the voiceless: nature and future generations

The problems raised thus far are general problems for deliberative institutions that arise in any domain of choice. They are not peculiar to the environment. However, environmental decisions raise very particular problems for democratic theory concerning the nature and possibility of representation over and above those discussed this far. The central problem is that for many of those affected by decisions, two central features of legitimisation – authorisation and presence – are absent. Indeed for non-humans and future generations there is no possibility of those conditions being met. Neither non-humans nor future generations can be directly present in decision making nor can they authorise others. Clearly representation necessarily cannot be authorised by either non-humans or future generations nor rendered accountable to them.[17] And the politics of presence that underlies much of recent moves in deliberative democracy is ill suited to include future generations and non-humans: neither can be directly present.

In the case of current non-humans it might be objected that this might not be entirely true. Something like an Alejandro solution is possible. Consider the success of John Muir's strategy of taking Roosevelt out into the landscapes he aimed to preserve. There is a sense in which one might say that the strategy consisted in nature being represented by itself. However, while there is certainly a case more generally for taking deliberation into the places which are the object of deliberation, the articulation of any non-human interests or values here remains a human affair. Any representation of nature for itself, even if it were possible, would necessarily be selective. Consider another Borges story, *The Aleph*, in which the main protagonist is given access to one of what are called Aleph points, 'points in space that containing all other points' (Borges, 1967, p. 119), points at which the whole of life is present. There are no such points. Muir took Roosevelt to particular selected places which were to represent others. There are other places one could start to a different effect. Consider the passage from Miguel quoted in Chapter 7.

> [Miguel] pointed out the stonework he had done on the floor and lower parts of the wall which were all made from flat stones found in the Sierra. I asked him if he had done this all by himself and he said 'Yes, and look, this is nature' ('Si, y mira, esto es la naturaleza'), and he pointed firmly at the stone carved wall, and he repeated this action by pointing first in the direction of the Sierra [national park] before pointing at the wall again. Then, stressed his point by saying: 'This is not nature, it is artificial (the Sierra) this (the wall) is nature' ('Eso no es la naturaleza, es artificial (the Sierra) esto (the wall) es la naturaleza').
>
> (Lund, 2005, p. 382)

Miguel's carved wall represents a different part of nature under a different description to the Andalusian park to which he gestures. The presence of one and the absence of the other in public deliberation matters to its legitimacy. Consider again the Yosemite landscapes to which Muir takes Roosevelt. They walked not just in the presence of nature but also the absence of humans. The Ahwahneechee Indians were driven from their lands in Yosemite by Major Savage's military expedition of 1851. They still live outside the park boundaries to this day. As we noted in Chapter 7, their absence had its own mark on the ecology of the park. The grass parklands through which Muir and Roosevelt walked were in part the result of the pastoral practices of the indigenous people. What part of the world is represented for public deliberation, the state park or the stone fireplace, is a human affair. So also is the description under which it is represented, as a 'wilderness' or as a 'depopulated cultural landscape'. The presence of non-human nature in deliberation about environmental choices requires human representation.

Neither authorisation nor presence is possible in the case of non-humans and future generations. That this is the case is in one sense not a problem awaiting solution – it could not be otherwise. The problem lies in the claims to legitimacy of those current humans who claim to speak on the behalf of future generations and non-humans in the absence of those sources of legitimation. The standard solutions offered to the problem is that current generations authorise or act as trustees on behalf of the interests of non-humans and future generations. Thus the state (Pigou, 1952, pp. 29–30),[18] representatives from the environmental lobby (Dobson, 1996), and citizens who have internalised those interests (Goodin, 1996)[19] have all been suggested as proxy representatives. The solution of proxy representation is not however without problems. These lie in part with the historical precedents of this form of representation. One gets something akin to the idea of interests of one group can be represented by those of others, an idea which used to justify the representation of women by husbands and servants by masters. It underpinned the Whig notion of virtual representation employed to limit the extension the suffrage and characterised thus by Burke:

> Virtual representation is that in which there is a communion of interests, and sympathy in feelings and desires between those who act in the name of any descriptions of people, and the people in whose name they act, though trustees are not actually chosen by them. This is virtual representation.
>
> (Burke, 1792, p. 293)

The historical uses of the concepts of incorporated interests and virtual representation cast a shadow of illegitimacy over their revitalisation in green political thought. Moreover, it might be argued, even granted their legitimacy, they are ill suited to deal with the representation of non-humans and future generations, since it is simply false to assert that their interests are identical to those of current generations of humans. Non-humans and future generations are rather like the Catholics in eighteenth-century Ireland whom Burke claimed had no virtual representation since none with the same interests is 'actually represented' in the political process (Burke, 1792, p. 293).

What response can be made for the virtual representation of non-humans and future generations? First, it might be suggested that the historical illegitimacy need not spill over to the representation of non-humans and future generations. It does not involve the relations of power and subordination that is involved in the illegitimate historical precedents, nor the failure to recognise the dignity of those denied direct representation. If individuals or groups can speak for themselves, then they should do so. However, where they cannot there is no loss in dignity nor assumed power relations in others speaking for them: the representation of infants through the adults who care for them, while not perfect, is not illegitimate. Hence there is nothing as such illegitimate about representing the interests of future humans and non-humans through current persons (Goodin, 1996).

What, however, of the objection that there is no identity of interests between current humans and future generations and non-humans? There are two broadly Kantian responses that might be employed here. The first response is one that appeals to the publicness condition on deliberation, reasons must be able to survive being made public. The publicness condition is taken to force participants to offer reasons that can withstand public justification and hence to appeal to general rather particular private interests. The point is one that has roots in Kant for whom every moral principle must meet 'the formal attribute of publicness': 'All actions affecting the rights of other human beings are wrong if their maxim is not compatible with their being made public' (Kant, 1793, Appendix, II). The test rules out those arguments from principles that appeal to self-interests where this conflicts with just concern with the interests of others, since the persuasiveness of such arguments could not survive publicity (Rawls, 1996, pp. 66–71; Elster, 1998a). Hence, reasons for action that appeal to wider constituencies of interest – including those of future generations and non-humans – are more likely to survive in public deliberation than they are in private market based methods for expressing preferences (Goodin, 1996, pp. 846–847; Jacobs, 1997).

Goodin develops this further and suggests that through such deliberation wider interests are internalised, to view democracy 'as a process in which we all come to internalize the interests of each other and indeed of the larger world around us' Goodin, 1996, p. 844). Through the internalisation of the interests of nature those interests can be virtually represented:

> Much though nature's interests may deserve to be enfranchised in their own right, that is simply impracticable. People, and people alone, can exercise the vote. The best we can hope for is that nature's interests will come to be internalized by a sufficient number of people with sufficient leverage in the political system for nature's interests to secure the protection they deserve.
> (Goodin, 1996, p. 844)

This thought of individuals thus representing the standpoint of all others might itself be taken to involve a second Kantian move of the kind to which Arendt appeals. Thus her account of 'representative thinking', like that of Goodin, appeals to the idea of deliberation as a process in which the outlooks and interests of others are internalised, and like Goodin she takes this not only to involve public deliberation, but also to use his phrase 'democratic deliberation within' (Goodin, 2000). Her account is worth quoting at length:

> Political thought is representative. I form an opinion by considering a given issue form different standpoints, by making present to mind the standpoint of those who are absent; that is, I represent them. The process of representation does not blindly adopt the actual views of those who stand somewhere else, and hence look upon the world from a different perspective; this is a question neither of empathy, as though I tried to be or feel like somebody else, nor of counting noses and joining a majority, but of being and think in my own identity where actually I am not. The more people's standpoint I have present in my mind while I am pondering a given issue, and the better I can imagine who I would feel and think if I were in their place, the stronger will be my capacity for representative thinking and the more valid my final conclusions, my opinions. (It is this capacity for an 'enlarged mentality' that enables men to

> judge; as such it was discovered by Kant in the first part of his *Critique of Judgement*, though he did not recognize the political and moral implications of his discovery.) The very process of opinion formation is determined by those in whose places somebody thinks and uses his own mind, and the only condition for this exertion of the imagination is disinterestedness, the liberation from one's own private interests. Hence, if I shun all company or am completely isolated while forming an opinion, I am not simply together with myself in the solitude of philosophical thought; I remain in the world of universal interdependence, where I can make myself the representative of everybody else.
>
> (Arendt, 1968, pp. 241–242)[20]

There are, however, clear difficulties in an appeal to an Arendtian and Kantian account of representative thinking in this context. The Kantian argument concerns the reliance of judgements on a wider community of judgement. The power of judgement is social in that it relies on the comparison of judgements with others in order to 'escape the illusion that arises from the ease of mistaking subjective and private conditions for objective ones, an illusion that would have prejudicial influence on the judgement' (Kant, 1987, pp. 293–294). This thought is important, but it won't extend to non-human animals. They lack the capacity for judgement and hence do not belong to that community to which judgement must test its soundness.

The idea of internalising interests, in the sense that Goodin makes it, has then to be independent of this Kantian line of thought. Rather, the argument needs to run that through the public use of reason, one escapes not just the narrowness of partial judgement but also the narrowness of private interests. Enlarged interests can survive public deliberation and develop a wider perception of what and whose interests count. However, this raises a second problem. While representative thought can take place in solitude, politics is not just about thought, but speech and representative speech is public and demands justification. On what grounds can we hold that you are able to speak for others? The issue remains as to what, in the absence of authorisation, accountability or shared identity, can legitimise any particular individual or group making public claims to speak on behalf of the interests of others. Goodin may be right here that the internalisation of interests is the best one can hope for in terms of representing nature or future generations, but the idea of internalising interests does not resolve the problems concerning the legitimacy of public representation of those without voice.

11.7 Speaking for nature?

What response can be made to a challenge of the legitimacy of a claim to speak for others? In the absence of authorisation, accountability and presence, the remaining source of legitimacy to claim to speak in such cases is epistemic. Those who claim to speak on behalf of those without voice do so by appeal to their having knowledge of the objective interests of those groups, often combined with special care for them. Thus, natural scientists, biologists and ecologists are often heard making special claims 'to speak on behalf of nature' where their claim to do so is founded upon their knowledge and interests. Environmental lobby groups make similar claims. However, as I noted at the outset of this chapter, such claims are also commonly disputed.

Two kinds of dispute are of particular significance here. First, there are new versions of a traditional normative debate in political theory about the proper descriptions under which the representation should take place. Is it as individuals or as members of particular groups or as bearers of particular identities? The animal liberation and welfare movements are individualist: it is individual sentient beings that have moral considerability and it as such that they should be represented in our decisions. Those involved in the environmental movement and nature conservation are concerned primarily with the conservation of biodiversity, species and habitats, and it is as members of particular species or as bearers of particular roles within an ecological systems that non-human nature is to be represented.[21] The dispute is normative. The debate is not primarily about particular knowledge claims as such, but rather about which knowledge claims are normatively relevant to the representation of nature – those concerning the welfare of individuals or those concerning the functioning of ecosystems and habitats. The conflict itself is in practice a real and important one. While it is often the case that what is good for habitats will also be good for individuals, there are a series of practical issues where the two come apart, for example the culling by conservation authorities of feral animals or non-native animals in order to preserve some particular habitat or the release by animal rights activists of caged animals into non-native habitats. The spokespersons for nature speak in different voices.

A second set of disputes arises with direct challenges to the epistemic legitimacy claimed by putative scientific spokespersons for nature. These are of particular significance when representatives of nature have too much voice rather than too little, for example, when they conflict with communities speaking with an already marginalised voice who are policed in and excluded from 'nature reserves' justified by natural scientists. Consider again the remarks of Miguel quoted earlier. They are typical of a number of disputes between the representatives of nature on the one hand and the communities they aim to police in nature's name on the other. The disputes are particularly evident in the export of nature reserves to the developing world, discussed in Chapters 7 and 8. They are at their most acute where nature is evoked to justify the exclusion of people from their homes.

Conflicts exist between international conservation bodies speaking on behalf of the interests of nature seeking to protect 'natural landscapes' and the socially marginalised groups whose lives and livelihoods depend on working within them. Places matter to such groups in ways that conflict with the goods defended by the representatives of nature. Such groups have particular local knowledge of place which gives them a distinct voice in its future different from that of the scientific expert who claims to speak on behalf of nature. The well-discussed arguments in political epistemology about whose knowledge claims count in environmental decision making is in part an argument about the legitimacy of representation founded purely on epistemic authority: who can claim to speak on behalf of others, where the only claims for legitimacy are knowledge claims, and authorisation, accountability or presence are impossible?

To raise these questions is not to offer any solutions to the specific problems of representing future generations and the non-human world. Indeed I suspect there are no solutions as such. However they do point to a need to shift away from one traditional justification of deliberative democracy. Recent accounts of deliberative democracy, especially those that have their roots in Habermas, define successful deliberation in terms of the convergence of judgements under ideal counterfactual conditions in which all have equal voice and 'no force except that of the better argument is exercised' (Habermas, 1975, p. 108).[22] However, it may be that there is no possibility for convergence even in ideal conditions. I say this not because of the common argument that for normative questions there are no truths of the kind found in science on which one could expect convergence through reasoned debate (Williams, 1985, ch. 8). I remain unconvinced by this argument (O'Neill, 2001b). Still less am I convinced by the post-structuralist reduction of truth to power (O'Neill, 1998a, ch. 7). I say it rather because, for reasons I have outlined in Chapter 8, there can exist conflicts between different goods themselves, which give reasons for scepticism about the possibility of convergence even under ideal conditions.

Values are plural and can conflict in ways in which no resolution is possible. Different human goods fostered within distinct practices by different groups may themselves be in conflict. Thus for example it is possible in some, although by no means all, contexts for good husbandry of the land within a marginal farming community to come into conflict with the aims of conserving biodiversity (O'Neill and Walsh, 2000). Such conflicts may be unresolvable. Conservation does not have some lexicographic priority in such cases over the value of community or the internal goods of husbandry. Nor are there simple trade-offs between commensurable values. There are real human goods that demand realisation but which conflict. Even under the ideal conditions hard and sometimes irresolvable choices between values that contingently conflict cannot be eliminated from ethical and political life. Moreover under the non-ideal conditions of actual deliberation in political and social life, the existence of consensus can be a sign of personal or structural power which is exercised to keep various voices and conflicts out of the realm of public discussion, rather than an indication of the exercise of the power of reasoned public conversation.[23]

In this context, the virtue of deliberative democracy may lie not in claims that it resolves conflicts but that it reveals them. The publicness condition forces participants, in at least some conditions, to justify their claims. Public deliberation in this respect differs from expressions of private preference in market behaviour which does not require justification. But publicness here is a virtue because it opens up space for contesting claims, not, as is suggested in some deliberative theories of democracy, because it is a condition for consensus. For that reason there is as much a need for dissensus conferences as consensus conferences. More generally it raises issues about the nature of the ideal of 'inclusion' which I touched upon in Chapter 9. The inclusion of different voices can take place under a variety of different terms of engagement. And under some terms of engagement, the result of 'inclusion' of persons might paradoxically be the silencing of the particular voices they are taken to represent. For example, there is a particular need for spaces in which social movements can maintain conversations that are oppositional to dominant free-market discourse of globalisation. The inclusion of movements in official processes of deliberation can in many contexts effectively silence that conversation. Given the real asymmetries that exist in power and voice there is a need for places where hidden conflicts are made explicit and marginal voices are articulated and heard.[24]

12 The political economy of deliberation

12.1 Deliberative democracy and political economy

I opened this book with the question as to the source of our environmental problems. Why is there in modern societies a persistent tendency to environmental damage? The answer that neoclassical economics gives to that question is that the source of our environmental problems lies in the absence of markets in environmental goods, their solution lies in extending market prices to ensure that the value of environmental goods is properly recorded. The shared assumption of both Austrian and neoclassical economics, that rational choice requires the use of market mechanisms and monetary measures, remains at the core of current debates in environmental economics. For reasons we have discussed in this final part of the book, the standard critical response to market solutions has been the shift from the market to the forum, to experiments in deliberative institutional frameworks for social choice. In place of the market picture of democracy as a procedure for aggregating and effectively meeting the given preferences of individuals, the deliberative theorist offers a model of democracy as a forum through which judgements and preferences are formed and altered through reasoned dialogue. In practice the revival of deliberative democracy has been expressed in the development of a variety of 'new' formal deliberative institutions which are often presented as experiments in deliberative democracy such as citizens' juries, citizens' panels, in-depth discussion groups, consensus conferences, round tables, and, more problematically, focus groups. At the same time at the level of decision-making tools there have been attempts to develop forms of multi-criteria decision aid that recognise the irreducibility of different dimensions of value involved in social choices. Where such multi-criteria decision aids have in the past tended to be technocratic and expert based, the recent move has been to find ways of integrating them into deliberative institutional contexts (De Marchi et al., 2000).

Deliberative theory has an egalitarian component. The power of voice must be roughly equal if deliberation is to meet the norms of communicative rationality. Deliberation requires both formal and substantive equality. Substantive equality is present to the extent that internal dialogue is not distorted by the distribution of power and resources between participants in dialogue: 'The participants are substantially equal in that the existing distribution of power and resources does not shape their chances to contribute to deliberation, nor does that distribution play an authoritative role in their deliberation' (J. Cohen, 1989, p. 23). The liberal ideal, which is largely shared by Habermasian models of deliberation, is of autonomous deliberative institutions protected from those patterns of power in the economy.[1] The sphere of deliberation must be protected from colonisation by the economic realm. However, whatever the defensibility of that liberal ideal, given the existence of large inequalities in economic and social power, the liberal ideal on its own terms becomes untenable. The structural conditions for deliberation are absent: formal equality in democratic rights is insufficient without a rough equality in social and economic power. The problem is part of a more general weakness in recent work on deliberative democracy in its avoidance of issues in political economy. The theories tend to a purely symbolic or cultural politics which fails to address the ways in which the structural imperatives of markets place constraints on the actual decisions of actors (O'Neill, 1998a, ch. 1).

Thus, to the extent these experiments in deliberation are presented as purely political solutions in the absence of changes to economic institutions there are two grounds for scepticism from an egalitarian direction. The first appeals to well-worked arguments against the purely formal equality involved in liberal institutions. While 'new' deliberative institutions are often presented as 'facilitating' 'inclusive' 'dialogue' between equals, dialogue takes place against the background of large asymmetries of social, institutional and economic power. A rough equality in economic and social power is a necessary condition for good deliberation, and such equality requires not just the formal status of equality in citizenship at the level of politics, or a purely symbolic acknowledgement of identity at the level of culture. It requires shifts away from current patterns of ownership and control of resources.

A purely political model of deliberation is also open to a second objection concerning the split between citizen and market actor it assumes. One version of this is that in Marx's 'On the Jewish Question' (Marx, 1974). In the perfected democratic constitution the individual lives in two worlds and two roles – as a citizen in the communal world of democracy and as a self-interested agent in the world of the market. Social emancipation requires the transformation of the economic realm of bringing communal citizens down to earth in their everyday life 'when the real, individual man re-absorbs in himself the abstract citizen'. The division that Marx criticises – between 'citizen' and 'market agent' – is assumed by many who advocate a deliberative response to environmental problems. Consider for example the distinction between the preferences an individual has in the role of consumer and those that he or she has as a citizen drawn by Sagoff (1988) in his influential *The Economy of the Earth*. As a consumer an individual expresses 'personal or self-regarding wants and interests', as a citizen an individual expresses 'judgements about what is right or good'. For Sagoff it is in the role of citizen that the individual deliberates about environmental goods. The mistake of market approaches to environmental problems is that they transform an issue that requires public deliberation by citizens into one to be resolved by consumer preferences. Sagoff's use of the distinction between citizen and consumer takes for granted the institutional contexts of different preference orders. That split is problematic. It is unclear just where and what the boundary of the political and the market is supposed to be: there are very few economic decisions that do not have environmental consequences. Within the economic sphere itself, to leave the allocation of most resources to the market is incompatible with the realisation of environmental goods. The market responds only to those preferences that can be articulated through acts of buying and selling. Hence the interests of the commercially inarticulate, both those who are contingently so – the poor – and those who are necessarily so – future generations and non-humans – cannot be adequately represented. Moreover, a competitive market economy is necessarily orientated towards the growth of capital, and hence is incompatible with a sustainable economy. The idea we can simply live in two worlds with the kind of preference schizophrenia that Sagoff assumes is untenable. Public deliberation needs to be taken into the economy itself.

This case for bringing deliberative institutions into the economic realm is, however, open to some major objections. Among the most significant of these is a version of Hayek's epistemological criticisms of planned economies which has recently been deployed against the possibility of a predominantly non-market economy coordinated by deliberative institutions (Hodgson, 1999, pp. 48–49) or more strongly against the use of deliberative responses to environmental problems as such (Pennington, 2001). Hayek's epistemic arguments for the market and against planning run in outline as follows. The market is presented as a solution to problems of ignorance that planned economies cannot solve. One central source of human ignorance is what Hayek calls 'the division of knowledge' in society. The central point here is not simply the dispersal of knowledge throughout different individuals and groups in society, but the nature of that knowledge. It includes practical or 'tacit' knowledge often embodied in practices and skills, and knowledge of particulars localised to a specific time and place. Such knowledge cannot be articulated in general propositional form. Hence it is not the kind of knowledge that could be passed onto a single planning agency. Hence no such agency could in principle possess all the knowledge required for coordination of the activities of different economic actors. The price mechanism in contrast does communicate between individuals that information that is required for the coordination of their economic activities, while allowing them to employ their particular knowledge (Hayek, 1949a, 1949b).[2]

This argument against central planning and in defence of markets has been redeployed against deliberative models of decision making thus: If there are forms of tacit or practical knowledge that cannot be passed on to a central planning system because they are not open to articulation, then neither will they be open to articulation in a deliberative setting. Hence, models of economic decision making that rely purely on deliberative institutions to coordinate activities will be open to some of the same objections as is made to the epistemic case against planning (Hodgson, 1999, pp. 48–49; cf. Pennington, 2001). The argument runs to the conclusion that any coordination of dispersed tacit knowledge requires mechanisms that 'rely to some significant extent on the market and the price mechanism' (Hodgson, 1999, p. 49). How compelling is this argument? In order to consider its power it is worth putting this argument in the context of Hayek's own debates with the tradition of thought that issued in what is now called ecological economics.

12.2 Hayek, epistemology and ecology

Hayek's epistemic criticisms of planning were directed not only at socialist planning but also at the tradition of thought that lies behind what is now called ecological economics (Martinez-Alier, 1987). Two assumptions which are central to that tradition are questioned by Hayek (1941, 1942–1944): first, that economics should be concerned with the ways in which economic institutions and relations are embedded within the physical world and have real physical preconditions which are a condition of their sustainability; second, that rational economic choices cannot be founded upon purely monetary valuations but require direct reference to their physical characteristics. Precursors of ecological economics such as Ostwald, Geddes, Soddy and Solvay are all objects of criticism, their work on energy units taken to exemplify 'scientistic objectivism' typical of an engineering mentality. Their objectivism that is exhibited in their belief in the desirability for *in natura* calculations in kind in economic choices as against calculations in monetary valuations (Hayek, 1942–1944, pp. 90 and 171).

For Hayek the ecological tradition exemplifies a 'scientistic objectivism' expressed in

> the characteristic and ever-recurrent demand for the substitution of *in natura* calculation for the 'artificial' calculation in terms of price or value, that is, of a calculation which takes explicit account of the objective properties of things.
>
> (Hayek, 1979, p. 170)

The belief in objectivism and *in natura* calculation for Hayek expresses an illusion about the scope of knowledge that is typified by the social engineer's belief in some purely technical optimal solution to social choices (Hayek, 1979, p. 170). In holding to the possibility of such an optimum the social engineer is a victim of the illusion of complete knowledge that underpins the project of socialist planning (Hayek, 1979, p. 173). It exhibits 'Cartesian rationalism', the belief in the omnipotence of reason and the corresponding failure of 'human reason rationally to comprehend its own limitations' (Hayek, 1979, p. 162). Cartesian rationalism in the social domain fails to acknowledge the division of knowledge in society, the dispersal of practical knowledge embodied in skills and know-how, particular to local time and place. Such knowledge cannot be articulated in propositional form and hence cannot in principle be passed on to a central planning body. The market in contrast resolves problems of ignorance by acting as a coordinating procedure which, through the price mechanism, distributes to different actors that information that is relevant for the coordination of their plans (Hayek, 1949a, 1949b). Given this view of the price system, to give up prices for calculation in kind is to give up a solution to the problem of ignorance for the illusion of the possibility of complete knowledge based in planning. There is no *in natura* alternative to the monetary measures.

How convincing are Hayek's epistemic criticisms of the underpinning of ecological economics? It is undoubtedly true that a number of precursors of ecological economics did hold on to the doctrine that there exists some set of physical energy units that allow us to plan in an ecological rational manner. And it may even be true that such a doctrine persists among some advocates of ecological economics. However, the doctrine is not entailed by the two central claims of the tradition that Hayek criticises nor by some of the figures in the tradition he opposes. Consider for example one of Hayek's central opponents in the debate, Otto Neurath.[3] The doctrine that Hayek criticises – that there are some purely physical units, like units of energy, independent of human use or belief that could be employed for planning – was rejected by Neurath and with it the technocratic idea that there is any 'optimum' solution to social problems. Neurath opposes 'what is called the "technocratic movement"' which assumes there exists

> one best solution with its 'optimum happiness', with its 'optimum population', with its 'optimum health', with its 'optimum working week', with its 'optimum productivity' or something else of this kind [and which] asks for a particular authority which should be exercised by technicians and other experts in selecting 'big plans'.
>
> (Neurath, 1942, pp. 426–427)

Neurath's criticism of technocratic planning moreover starts from epistemic assumptions shared with Hayek. For Neurath like Hayek, proper rationalism recognises the boundaries to the power of reason in arriving at decisions: 'Rationalism sees its chief triumph in the clear recognition of the limits of actual insight' (Neurath, 1913, p. 8). It is a mark of what he calls 'pseudorationalism' to believe that there exist technical rules of choice that determine optimal answers to all decisions.

However, in the work of Neurath the argument against Cartesian rationalism or pseudorationalism is employed in a very different direction. It is employed against any algorithimic conception of rational choice that assumes there exists some single unit of calculation, monetary or non-monetary, which can be employed to reduce decision making to a matter of calculation. Hence for example the following passage on choosing between alternative sources of energy:

> The question might arise, should one protect coal mines or put greater strain on men? The answer depends for example on whether one thinks that hydraulic power may be sufficiently developed or that solar heat might come to be better used, etc. If one believes the latter, one may 'spend' coal more freely and will hardly waste human effort where coal can be used. If however one is afraid that when one generation uses too much coal thousands will freeze to death in the future, one might use more human power and save coal. Such and many other non-technical matters determine the choice of a technically calculable plan . . . we can see no possibility of reducing the production

plan to some kind of unit and then to compare the various plans in terms of such units.

(Neurath, 1928, p. 263)

Given a choice between alternative sources of energy – say coal and hydraulic power or solar energy – a variety of ethical and political judgements for example about intergenerational equity and the distribution of risks come into play. One cannot make such choices through a purely technical procedure employing some single unit either monetary or non-monetary.

However, while Neurath's conclusions differ from those of Hayek, his starting point is, as Neurath himself noted, similar. Neurath's criticisms of pseudo-rationalism parallel those of Hayek against rationalism. Both reject the illusion of complete knowledge on which technocratic planning is based (Neurath, 1945).[4] Neurath is of course not alone in sharing this perspective on the socialist side of the debate. Hayek has been taken seriously by many socialists not only because there is power in his epistemic criticisms of central planning but also because they share common ground with epistemic arguments offered within the socialist tradition. Hayek's epistemic criticism of centrally planned economies have their counterpart within the history of debates about socialist planning as an argument for democratic and decentralised decision making and for a proper appreciation of the limits of abstract technical expertise. There exists a long tradition of associational and libertarian models of socialism that has had argued against the possibility of an economy centralised on Fabian or Bolshevik lines (Schecter, 1994).[5] Similar epistemic themes concerning the importance of local ecological knowledge embodied in particular human practices, the inescapability of social choice in conditions of radical ignorance and the limits of scientific predictability are central to the ecological tradition. These concerns underpin the need for cultural and institutional pluralism, the precautionary principle and the importance of deliberative institutions which counter the necessary role of the scientific expert. The central question which divides Hayek's position from his socialist and ecological opponents is how far specifically market institutions are either necessary or sufficient to coordinate practical knowledge. Are there alternatives to market solutions to the coordination of knowledge? The recent redeployment of Hayek's arguments against deliberative models of economic life have brought that question back to the fore. The conclusion that Hodgson wants to draw is that any coordination of dispersed tacit knowledge requires mechanisms that 'rely to some significant extent on the market and the price mechanism' (Hodgson, 1999, p. 49).

Does the coordination of dispersed knowledge require significant reliance on the market mechanism? Oddly, some of Hodgson's and Hayek's own arguments can point in the opposite direction. When Hodgson wants to stress the role of practical knowledge, he turns, as did Hayek in places, to Polanyi's discussion of tacit knowledge in science: 'As Polanyi explained, all scientific advances and technological innovations are bound up with tacit knowledge. They rely on accumulated skills and habits, implanted in individuals and institutions' (Hodgson, 1999, p. 49). In this I think Hodgson is entirely right (O'Neill, 1998a, pp. 150ff.).

However, the example is odd if the point being made is the necessity of markets and prices to coordinate tacit knowledge. The coordination of the scientific knowledge, both tacit and articulated, is, as Neurath stresses, one of the achievements of the modern world. What he calls the orchestration of the sciences is taken to illustrate 'how much unity of action can result, without any kind of authoritative integration' (Neurath, 1946, p. 230). Science is integrated without any pyramidic organisation. The model of orchestration of the kind one finds in the sciences is one that reflects Neurath's vision of planning, in particular in his later writings, as a possible 'future order based not on "state pyramidism", if this term may be accepted, but on "overlapping institutions", which do not coincide with any "hierarchic" world pattern' (Neurath, 1943, p. 149). And whatever the plausibility of that vision there is an important point here – the scientific community is an example of an international non-market order within existing economies. It offers one of the many examples of non-market coordination of knowledge and activities within existing market societies. Moreover, the danger in recent developments in intellectual property rights regimes is that introducing market mechanisms into the production and coordination of knowledge will slow rather than increase rate of innovation. The conflicts between proprietary control of knowledge and the need for its public availability is about the control of knowledge crucial to innovation (O'Neill, 1998a, ch. 11).

More generally, the coordination of knowledge, tacit and articulated, is a ubiquitous problem that exists at all points in the social and economic order. As knowledge and skills have come to the centre of economic life in the modern world, the management of the distribution of unarticulated knowledge throughout organisations of various kinds without its loss has become an increasingly important task. The market is not a requirement for coordination and as the case of science shows, in some cases can become a hindrance. There exist at a variety of levels of coordination throughout the economy in different institutional forms. Moreover, our knowledge of those coordinating mechanisms is itself often practical and unarticulated: they are embodied in a variety of practices which we cannot completely survey. All institutions are built on forms of habitual forms of behaviour that embody practical knowledge.

To defend these claims is to accept what Hodgson calls an 'impurity principle' (Hodgson, 1999, p. 126) in so far as this is understood as a claim for institutional pluralism. There is clear common ground between Hodgson and Neurath at a general level on the need for such pluralism. Thus Neurath defends '*economic tolerance* that can support several non-capitalist forms of economy simultaneously' (Neurath, 1920b, emphasis in the original; cf. 1920a, section 22). There are good epistemological reasons for such tolerance – different forms of knowledge are embodied in different practices. Moreover there are reasons to do with the nature of human goods involved. For example, recognition of the worth of a scientific achievement, health, and love cannot given their different natures be distributed by identical institutional principles: the first follows merit, the second need, the third the fortunes of kinship and affection. Hence the opposition to the distribution of many goods according to market principles. The point is one that Neurath develops against the use of money as a universal measure of value. Money is not a universal measure of value, but an element in a particular institution involving particular social relations. Similarly concepts like 'cost', 'profit' and 'investment', all have specific meanings within an institutional setting. The problems arise when one 'transfers' those concepts from one institutional setting to others in which they are no longer appropriate.

> To regard money as a historically given institution does not involve any objection to its use – though there may be such objections – but an objection to the application of arguments, valid in the field of higher bookkeeping, to the analysis of social problems and human happiness in general.
>
> (Neurath, 1944, p. 39)

Neurath objects to the uniformity of practices that markets themselves can bring in their wake.

These claims about the uniformity engendered by market relations have implications for Hayek's own epistemic defence of the market. As I have argued elsewhere (O'Neill, 1998a, ch. 10) the market stands in a much more problematic relation to local and practical knowledge than Hayek assumes. Far from fostering the existence of practical and local knowledge, the spread of markets often appears to do the opposite: the growth of global markets is associated with the disappearance of knowledge that is local and practical, and the growth of abstract codifiable information. Hence, there is then something at the least paradoxical about Hayek's epistemic defence of the market, for the market as a mode of coordination appears to foster forms of abstract codifiable knowledge at the expense of knowledge that is local and practical. The knowledge of weak and marginal actors in markets, such as peasant and marginalised indigenous communities, tends to be lost to those who hold market power. The epistemic value of knowledge claims bear no direct relation to their market value (Martinez-Alier, 1997, p. 194). Local and often unarticulated knowledge of soil conditions and crop varieties that have considerable value for the long-term sustainability of agriculture has no value in markets and hence is always liable to loss when it comes into contact with oil-based agricultural technologies of those who do have market power. The undermining of local practical knowledge in market economies has also been exacerbated by the global nature of both markets and large corporate actors who require knowledge that is transferable across different cultures and contexts and hence abstract and codifiable: 'quantification is well suited for communication that goes beyond the boundaries of locality and community' (Porter, 1995, p. ix). Finally, the demand for commensurability and calculability runs against the defence of local and practical knowledge. This is not just a theoretical problem but one with real institutional embodiments. The market encourages a spirit of calculability – of rationalisation in Weber's sense. That spirit is the starting point for the algorithmic account of practical reason which requires explicit common measures for rational choice and fails to acknowledge the existence of choice founded upon practical judgement. More generally it is not amicable to forms of knowledge that are practical, local and uncodifiable. The forms of rationalism that Hayek criticises have roots in the market institutions he defends.

Hayek's defence of market institutions is unsatisfactory. However, as I have argued elsewhere, the failures in his positive account of markets does not mean that his negative case against centralised planning fail (O'Neill, 1998a). Neither do they as such undermine the redeployment of those arguments against bringing public deliberation into economic life. Where then does all this leave the Hayekian arguments against deliberative institutions in economic and social life? The answer depends on the assumptions that are made about the nature and scope of such institutions. If one assumes a strongly intellectualist account of the nature of deliberation according to which deliberation is understood to consist of in the direct exchange of arguments about the truth or rightness of propositions, then part of the Hayekian criticism has to be granted. However, we have already given reasons in Chapter 10 to be critical of such intellectualist accounts of deliberation in discussing the role of testimony in deliberation. Any deliberative forum itself relies on a background of trust in the testimony of others. Judgement is not passed directly on the knowledge claims, but the credibility and trustworthiness of those who make them.

More generally, any deliberative institutions rely for their operation on institutional arrangements and habitual forms of behaviour that are not themselves direct objects of deliberation. Indeed the very process of deliberation itself relies on practical skills and know-how that themselves cannot be fully articulated in propositional form. Likewise there are limits in the scope of deliberation. Coordination of activity is also a matter of institutional habits and skills. That this is the case does carry problems for a strong interpretation of Marx's claim that our social relations should be subordinated to conscious communal control (Marx, 1973, p. 162). There are epistemological limits to any such project. However, none of this entails that public deliberation or participatory planning ought to have no central role in political and economic life. While the Marxian claim that one can bring all social life under conscious control might be a mistake, it would be equally mistaken, as Hayekians and conservatives more generally often do, to thereby reject the whole project of rational reflection on the direction of social relations. The limits of deliberation can be granted without rejecting the role of deliberative modes of coordination for purely market-based modes of coordination. Again the scientific community is case in point. The forms of local coordination involved in scientific communities are deliberative in nature.

12.3 Pluralism, markets and the present of the future

The objections to Hayek outlined in section 12.2 are not ones with which critics of non-market deliberative models of socialism, like Hodgson, would necessarily disagree. Thus Hodgson's defence of institutional pluralism is aimed as much against the free-market proposals of Hayek as it is against the proposals of non-market socialists. And again there are versions of socialism that are open to that objection. Indeed a powerful objection to the position which Marx (1974) defends in 'On the Jewish Question' is the institutional monism it involves, a monism which has been a feature of much socialist thinking during the twentieth century. Whatever the specific problems with market economies, the assumption that the division between the private person and the public citizen is one that should simply be overcome is problematic (Keat, 1981). However, for reasons just noted, the criticism of institutional monism also cuts in the opposite direction. The recognition of the need for institutional pluralism has been point central to ecological criticism of the market economy. Ecological knowledge is embodied in particular practices and traditions that cannot be articulated and which globalised markets undermine (Martinez-Alier, 1997, 2002; O'Neill, 1998a, ch. 10). The argument here appeals to an argument for pluralism within the anti-capitalist tradition that recent resistance to neo-liberal globalisation has brought back to the fore.

Nor is it the case that any argument for institutional pluralism is per se an argument for the necessity of market relations. Thus, for example, while Hodgson and Neurath concur on the general principle of institutional pluralism, they differ on the question of whether markets form a necessary part of any plural institutional order. Neurath does not object just to the spread of market relations beyond their proper scope, but to a market economy as such. His own pluralism was basically one of non-market orders (O'Neill, 2003a). My own judgement is that Neurath is right to retain some scepticism about the claims made for the necessity of a market economy in the modern world, although this is not a judgement I have defended here.

The defence of institutional pluralism is one that in current debates does, however, need stressing. There is a tendency among socialists and other opponents of modern capitalist society to think in terms of some single alternative model of social relations. There is a pluralism of non-market forms of social and economic relations – gift, public networks of recognition, scientific communities, kin relations, cooperatives etc. It is a mistake in considering the future of post-capitalist societies after the failure of centralised state economies to look for *the* new model. Given the knowledge that is embodied in human traditions and practices there is are good epistemological grounds for scepticism about the attempt to reorganise all human activities on the basis of some general set of rational principles. As Hayek and Neurath both note, reason sees its maturity in the recognition of its own limits.[6]

Conversely, there is a need to recognise the significance of non-market orders that exist within existing society, to defend these against market encroachment while remaining critical of their own internal forms of power. Neo-liberal forms of globalisation have seen the geographical spread of commercial norms into communities which retain non-market economic relations employing common property, mutual aid and gift, for whom markets previously existed on the margins of their economic life. At the same time in advanced market economies, there has been an expansion in the domains of goods and relations falling under market norms. Domains that were previously outside market norms such as public science, access to information, control of agricultural and natural biodiversity, and bodily integrity are being redefined as potential spheres of market exchange that come under new property rights regimes. As we saw in Chapter 1, environmental goods are increasingly coming under the norms of market exchange in both these senses. A link exists between current resistance to spread of market norms and property rights – keeping knowledge free, sustaining islands of cooperative activity, maintaining access to environmental goods, – and possibility of a future non-capitalist social order. The significance of the recent anti-capitalist movement against neoliberal globalisation has been the defence of non-market institutional and cultural forms against the spread of market norms. Those of us who are committed to the possibility of an alternative to existing society need not just to recreate utopian visions but build upon existing non-market orders – to remind ourselves of the existence of social relations and networks both local and global that have developed outside market norms. The future of post-capitalist societies is not ultimately compromised by possibility of feasible alternatives to the capitalist economy, although the unattractiveness of the putative alternatives in Eastern Europe and Asia that went under the title of socialism has left its legacy. The problems come increasingly from a form of social lock-in – that the capitalist market order survives because it becomes so tied into all human relationships that construction of an alternative becomes increasingly difficult to build or even imagine. In that context the defence of existing non-market orders becomes part of the project of maintaining the possibility of an alternative.[7]

Notes

1 Markets and the environment

1 I defend the project of resistance over boundary maintenance in the postscript to O'Neill (1998a). For an excellent general discussion of the problems of market boundaries which touches on the environmental case see Keat (1993, 2000, ch. 3). See also Walzer (1983, ch. 4); Anderson (1990, 1993); Radin (1996). For some sceptical remarks about the position I defend about environmental goods see Keat (1997) and Miller (1999).

2 Bava Mahalia (1994) 'Letter from a Tribal Village', *Lokayan Bulletin*, 11(2/3), Sept–Dec.

3 The term is that of Raz (1986). For discussions of the concept of incommensurability see Chang (1997).

4 The Levels themselves are a large wet grassland system designated a site of special scientific interest (SSSI) with important plants and invertebrates associated especially with the ditch systems. The respondents, drawn from residents, visitors and non-visitors to the Levels, were asked first if they were willing to continue to pay towards the WES in the Pevensey Levels. Those who were willing were then asked the maximum they would be willing to pay more for the WES – twice as much, three times as much and so on.

> Recall that the majority of households pay a few pence a year for the Wildlife Enhancement Scheme in the Pevensey Levels. Bearing in mind that there are many worthwhile nature conservation programmes in England which you might wish to support, what is the MAXIMUM your household would be willing to pay for the Wildlife Enhancement Scheme in the Pevensey Levels compared with today? Twice as much? Three times as much? Four times as much?
>
> (Willis et al., 1995, p. 4)

There are a number of internal methodological problems with that question and many contingent valuation practitioners would be rightly unhappy with the formulation of the willingness to pay question. However, many of the subsequent protests raise problems independent of any particular formulation of the question. They point to problems inherent to attempts to price the environment – and they echo some of the same problems illustrated by the letter on the Sardar Sarovar Dam: constitutive incommensurabilities; intergenerational community; competing property rights; problems of equity; problems of reason-blindness.

5 I discuss this in more detail in Part II of this book.

6 The general utilitarian idea that there is a measure of welfare gains and losses need not appeal directly to monetary measures. Consider the following comment from the cost-benefit analyst for the Narmada river dams, who, after allowing that suffering cannot be given a monetary value, continues to assume that there is a sum that can be done:

Then there was the question of the oustees – the trauma of the oustees. Now this was also given. Earlier we had not mentioned that there is indirect loss, but mind there is indirect benefit also. As I pointed out in the report, benefit is in the sense that if a person is uprooted from a place he suffers a trauma because of displacement, but at the same time, there are other people who gain because the water comes to them . . . In fact the beneficiaries are more than the sufferers. Therefore the quantum should at least be equal.

(Alveres and Billorey, 1988, p. 88)

7 Consider the following responses raised in the in-depth discussion groups to the contingent valuation question on Pevensey:

Sally: But I'll be quite honest, when I was asked that question, because of the financial situation my husband and myself are in, my answer was no, I couldn't afford to give *anything* extra because at the moment we're stretched to our limits. Because of losing jobs and that sort of thing. So, I then missed the bidding, which I was quite pleased about.
Susan: But the trouble is that would be, that could be misinterpreted couldn't it?
Sally: Absolutely. Absolutely. That's right.
Susan: As if you don't want it, you don't care (. . .) But I think, when we filled in that form I feel quite strongly that if we'd known a bit more, and the influence that perhaps the questions and our answers could have. Like when you said you couldn't afford to pay more because of your particular circumstance and if there was no facility to say that, they might get the wrong impression. And that, that fills me with horror to think we all might have been unemployed and we all couldn't have afforded to pay even if we wanted to.
David: That's right. That crossed my mind . . .

. . . . . .

Susan: . . . We might *all* have had a reason why we at this moment could not think in terms of paying and that, that bothers me. That bothers me. If it meant the future of the Levels depended on our particular circumstances.

(Burgess et al., 1995, p. 43)

8 For this reason the process, far from being rational, can appear rather whimsical. The point is made in the discussions of the respondents:

if it goes out of fashion, it's in danger all the time, isn't it? If the price drops, nobody's going to be interested. That aspect, that way of thinking, is not really on is it? You can't put it on the Stock Market really. It's our very existence. It's our future.

(Burgess et al., 1995, p. 45)

The problem of the reason-blindness in cost-benefit analysis is particularly evident in this issue of uninformed preferences. Consider the following worries that non-local non-visitors to the Levels had with the bidding questions:

I said to the interviewer, perhaps I should read up on Pevensey. She said no, you don't need to do that, that you needed if you like uninformed opinion to look at in a new light . . . So it depends on *your* approach. If you wanted just the public reaction then that's fair enough. But if you wanted a more informed reply, I think people have got to be put on their guard to think about it and to study what the objectives are . . .

without all the information you can't get the real answer . . .

It is pretty difficult to have real passionate feelings about something you know nothing about.

> It's not my area. I thought that's a question that cannot be answered. I didn't answer it. Well, I needed to know a lot more.
>
> (Burgess et al., 1995, pp. 72–74)

The responses here are quite proper ones. It is not elitist or paternalist to demand informed preferences: it is what is demanded of rational procedure for dealing with the problems, and it is not just 'elites' that have that knowledge. Information is required because it is an issue of argument about the merits of public projects, not about the aggregation of any preference, uninformed with informed.

9

> For a concept of the understanding, which contains the general rule, must be supplemented by an act of judgement whereby the practitioner distinguishes instance where the rule applies from those where it does not. And since rules cannot be in turn provided on every occasion to direct the judgement in subsuming each instance under a previous rule (for that would involve an infinite regress), theoreticians will be found who can never in all their lives become practical, since they lack judgement.
>
> (Kant, 1793, p. 61)

10 I develop these points and other constraints of reason at length in O'Neill (1998a, ch. 9).
11 I criticise that rationale in O'Neill (1998a, ch. 3).
12 My arguments in the following owe much to Wolff (2003).
13 Correspondingly some in-kind compensations may be acceptable where monetary compensation is not. Thus, for example, people will sometimes accept a public good as compensation for a loss of a public good, where no private monetary compensation will be acceptable. For discussions see Claro (2003); Field et al. (1996); Frey et al. (1996); Kunreuther and Easterling (1996); Mansfield et al. (1998).
14 An earlier version of this chapter appeared in *Economic and Political Weekly*, 36, 2001, pp. 1865–1873. I would like to thank the editors and publisher for their permission to use the material for this book.

2 Managing without prices

1 For useful discussions see the following: on hedonic pricing; Palmquist (1991) and Freeman (1995); on travel cost methods see Bockstael (1995); on contingent valuations see NOAA (1990); Fischoff, (1991); Loomis et al. (1993); Bishop et al. (1995); Pearce and Moran (1995, ch. 5); Bateman and Willis (1999).
2 It is worth noting here that utilitarianism *per se* does not commit one either to the use of monetary valuation or indeed to policy making procedures that aim directly at maximising total welfare. The preference utilitarian can recognise that there are some preferences that cannot be captured by monetary measures. Indirect forms of utilitarianism can recognise that there are good utilitarian reasons for not employing utilitarian decision procedures in public policy. The indirect utilitarian distinguishes between the criterion by which an action or policy is judged to be right and the decision-making method that is used, on a particular occasion, to decide which action or policy to adopt. While maximisation of welfare might be what makes a policy right, it need not follow that using decision making procedures that aims directly at the end will produce the right results. For example, they might result in calculative instrumental sensibilities that are not conducive to the kind of public social relations that best realise human well-being.
3 An earlier version of this chapter appeared as an article in *Ambio* 26, 1997, pp. 546–550. I would like to thank the editors and publishers for their permission to use the material in this book.

3 Property, care and environment

1 To say they are diffused and customary is not to say they are not well defined. The confusion of these concepts is a source of one mistaken source of neoclassical criticism of property systems that do not involve full liberal rights.

2 Here and later in the chapter I draw on interviews conducted in 1996 with thirty-seven farmers in three dales in the Craven district of North Yorkshire which lies just within the boundaries of the National Park. These semi-structured interviews lasted on average two hours. They were undertaken under a contract from English Nature to explore farmers' perceptions of nature conservation schemes in the Yorkshire Dales (Walsh et al., 1996; Walsh, 1997; O'Neill and Walsh, 2000).

3 Aristotle defends a system of private ownership and common use. Property is a condition for the exercise of the virtue of generosity. He also defends the view that property in land should not be alienable either through exchange or bequest. This view was a major influence on the civic humanist criticisms of the mobilisation of land by commercial society which is discussed further in section 5.2.

4

> In not asking for or expecting any payment of money these donors signified their belief in the willingness of other men to act altruistically in the future, and to combine together to make a gift freely available should they have a need for it. . . . As individuals they were, it may be said, taking part in the creation of a greater good transcending the good of self-love. To 'love' themselves they recognised the need to 'love' strangers. By contrast, one of the functions of atomistic private market systems is to free men from any sense of obligation to or for other men.
>
> (Titmuss, 1970, p. 239)

I discuss the relationships further in O'Neill (1992, 1998b).

5 Light (2000) has independently developed a related distinction between public goods and publicly provided goods, where the latter refer to goods that *ought* to be provided *as if* they were public goods.

6 Hardin comments: 'it is now clear to me that the title for my original contribution should have been *The Tragedy of the* Unmanaged *Commons* . . . I can understand how I might have misled others' (Baden and Noonan, 1998, p. xvii). This characterisation is still I think misleading. The tragedy is not about common property regimes. For a discussion see Aguilera-Klink (1994).

7 I discuss Locke's views in more detail in Chapter 7. For a discussion of the implications of Locke for recent discussions of property rights see Oksanen (1997).

8 For similar observations on the attitudes of farmers in the United States see Nassauer (1988, 1989).

9 My thanks to Tim O'Riordan for pointing out this passage for me and for some stimulating conversations about it.

10 An earlier version of this chapter appeared *Environment and Planning C: Government and Policy*, 2001, 19, pp. 483–500. I would like to thank Pion Limited, London for permission to reuse the material for this book. I would like to acknowledge the support of the Arts and Humanities Research Board in writing this article.

4 Public choice, institutional economics, environmental goods

1 For surveys see Mueller (2003) and Dunleavy (1991). The classic texts in the Virginian School of public choice theory are Buchanan and Tullock (1962) and Buchanan (1975). Typical of the Chicago School of public choice theory is Becker (1976). The Chicago version is much more closely tied to the extension of neoclassical theory to new domains than is the Virginian version which has taken an increasingly Austrian turn: consider the influence of Hayek on Buchanan's later work – see, for example, Buchanan (1986a) passim.

2 Compare Mueller: 'Public choice can be defined as the economic study of non-market decision making, or simply the application of economics to politics' (Mueller, 2003, p. 1).
3 See also Niskanen (1971). Downs (1967) defends a different version of a public choice account of bureaucracy, which includes altruistic motivations alongside narrower egoistic ones.
4 The major text defending this position is Downs (1957). An earlier classic attempt to apply economic models of behaviour to political actors, but with a more sceptical view of consumer sovereignty, is Schumpeter (1987).
5 Flyvbjerg et al. (2002) themselves do not endorse any public choice theory but rather appeal to a more general set of political explanations.
6 Buchanan attempts to show that the theorem still has relevance in conditions of imperfect competition (Buchanan, 1969) and, under an Austrian reinterpretation, that transaction costs are not relevant to its truth (Buchanan, 1986b).
7 I develop a version of this criticism of Sagoff in a Marxian idiom in O'Neill (1993a, ch. 10).
8 For a powerful defence of perfectionist liberalism see Raz (1986). I defend the compatibility of perfectionism and pluralism in O'Neill (1995b). It needs to be noted here that Mill's perspective on role of institutions on individuals' preferences sits uneasily with the psychologism he defend in Mill (1947, Book 6). However, it is the psychologism that needs to go, not the institutionalism.
9 See Polanyi (1957a, 1957b). For a discussion of the influence of Aristotle on Marx and Polanyi see Meikle (1979, 1985).
10 I include here not only J. S. Mill, but also the work of Hume and Smith. Compare, for example, the views about the role habit plays in the formation of dispositions of character in Hume – see 'Of the Origin of Government', 'The Sceptic' and 'Of Interest' (Hume, 1985) – and in Smith (1981, V.i), with those developed by Veblen (1919).
11 For a discussion see Polanyi (1957c), Meikle (1979) and O'Neill (1992, 1998b).
12 For an early survey see Dearlove (1989).
13 For classic discussion of this familiar point which is of particular relevance for the discussion here, see Taylor (1971).
14 See, for example, Buchanan and Tullock (1962, ch. 9).
15 Hence Hume's remarks about the incompatibility of absolute monarchy and commerce:

> Commerce, therefore, in my opinion, is apt to decay in absolute governments, not because it is there less secure, but because it is less honourable. A subordination of ranks is absolutely necessary to the support of monarchy. Birth, titles, and place, must be honoured above industry and riches. And while these notions prevail, all the considerable traders will be tempted to throw up their commerce, in order to purchase some of those employments, to which privileges and honours are annexed.
> (Hume, 1985, p. 93, emphases in the original)

See also Aristotle's remarks on the effects of wealth on character in Aristotle's *Rhetoric* book (Aristotle, 1946, II, ch. 16).
16 I owe this last point to Andrew Collier:

> if 'rational economic agent' is defined to mean 'person who pursues monetary gain in preference to all other aims' its corresponding term in ordinary English is not its homonym, but 'moneygrubber'.
> (Collier, 1990, p. 118)

17 As Hirschman goes on to say, it is perhaps true that:

> the success of Olson's book owes something to its having been contradicted by the subsequently evolving events. Once the latter had run their course, the many people

> who found them deeply upsetting could go back to The Logic of Collective Action and find in it good and reassuring reasons why those collective actions of the sixties should never have happened in the first place.
>
> (Hirschman, 1985, p. 79)

See also the problems that rational choice Marxism has with the explanation of collective class action.

18 In making that assumption they are the inheritors of the views of Smith: Smith was never an anti-socialist thinker – there was when he wrote no significant socialist movement to oppose. He was an anti-associationalist thinker: professional associations, trade associations and guilds are conspiracies against the public, concerned with pursuit of particular sectional interests (Smith, 1981, I.x, and passim).

19 I draw here on the distinction MacIntyre makes between practices and institutions (MacIntyre, 1985, ch. 14). To employ the language MacIntyre uses would, however, be misleading here, since he uses the term 'institution' in a very particular and idiosyncratic sense.

20 The point is made in a variety of ways, and by appeals to a number of authorities. Sometimes it is made in economic terms: that ethical constraints are a scarce resource, the use of which should be minimised (Buchanan and Tullock, 1962, pp. 27ff.). At others, in terms of justice, that the 'immoral' or 'egoistic' should not be allowed to gain 'unfair' advantage from his or her fellows (ibid. 302–306). Often it is simply invoked as a principle of institutional design that represents an inherited wisdom (Buchanan, 1978, pp. 17–18). In the final of these, Buchanan calls upon the authority of Mill's *Consideration on Representative Government* – 'the very principle of constitutional government requires it to be assumed that political power will be abused to promote the particular purposes of the holder' – conveniently forgetting the main thesis of that book concerning the educative and corrupting effects of institutions on the individual. Reference is also sometimes made to Hume:

> Political writers have established it as a maxim, that, in contriving any system of government, and fixing the several checks and controls of the constitution, every man ought to be supposed to be *knave*, and to have no other end, in all his actions, than private interest.
>
> (Hume, 1985, p. 42)

Hume goes on to say that 'it appears somewhat strange, that a maxim should be true in *politics*, which is false in *fact*' (ibid. pp. 42–43). While Hume does believe that 'avarice, or the desire of gain, is a universal passion which operates at all times, in all places, and upon all persons' (Hume, 1985, p. 113), his point about knavery in politics is very different from its public choice counterpart. It is one that applies solely to the political and concerns the behaviour of men when they act in parties, such that the countervailing check of honour is absent.

21 Hence, there is this to be said for rule by lot: that unlike modern elective oligarchy, it selects those who do not have a desire to rule (see Burnheim, 1985).

22 I begin to address these questions in O'Neill (1993b).

23 See also in this regard the rediscovery of associational models of socialism (Yeo, 1987; Hirst, 1989; Martell, 1992; O'Neill, 2003a).

24 This chapter draws on material that appeared in 'Public Goods, Environmental Goods, Institutional Economics', *Environmental Politics*, 1995, 4, pp. 197–218 (http://www.tandf.co.uk).

5 Time, narrative and environmental politics

1 For a detailed and more technical discussion of the concept of separability and different dimensions over which separability can be considered see Broome (1991).
2 I discuss these problematic features of the new world wilderness model in more detail in Chapter 7.
3 In that debate the role of narrative is raised most clearly by MacIntyre (1985, ch. 15). I discuss the issues raised here in more detail in O'Neill (1993a, ch. 3).
4 The source of Debord's views here Lukács' account of the transformation of time within the production process in modern capitalism: 'time sheds its qualitative, variable, flowing nature; it freezes into an exactly delimited, quantifiable continuum filled with quantifiable "things"' (Lukács, 1968, p. 90). For Debord, this transformation of time has now extended into all aspects of life.
5 The 'temptation to give credence to grand narrative of the decline of the grand narrative' (Lyotard, 1992, p. 29) has never been resisted. Moreover the narrative told about the rise and fall of modernity is itself implausible once any empirical detail is added.
6 This chapter draws on material from my contribution 'Self, Time and Separability', in B. Harrison (ed.) *Self and Future Generations* (Cambridge: White Horse Press, 1999). I would like to thank the publishers for permission to reuse this material here.

6 Sustainability

1 The classic example is Golding (1972).
2 Thus many problems such as the safe storage of nuclear waste are pressing because, whatever the technical feasibility of solutions that are offered, any technical solution that requires long-term maintenance also requires the social and cultural conditions which are presuppositions for the existence of the skills and knowledge necessary for their implementation. The problems will be most pressing where social disruptions are such that these conditions no longer exist.
3 For a development of this point see de-Shalit (1995).
4 Cited in Rawls (1972, pp. 290ff).
5 Norman Geras uses the passage from which this quotation is taken as evidence for the view that Marx does appeal to principles of justice even if he did not think he did (Geras, 1986, pp. 45ff.).
6 For an excellent discussion of different senses of exploitation, and in particular between neutral and 'pejorative' senses see Wood (1995).
7 In much of the standard economic literature on the environment welfare is defined in the standard way as preference satisfaction, and sustainability is defined in those terms. David Pearce (1993, p. 48) for example defines sustainability in terms of 'maintaining the level of human well-being' where well-being is defined in terms of preference satisfaction. However, 'the problem of entrenched deprivation' (Sen, 1992a, p. 55) suggests that preference satisfaction should not form the proper starting point for sustainability. The problem is a particular version of that of adaptive preferences. Individuals' preferences are formed relative to what they believe they can get. One response to believing that a good is impossible is to give up desiring it. In some cases the response displays a Stoic form of reasoning: cut your desires to the reality one faces. The consequence is that preference satisfaction cannot be an adequate metric for equality or sustainability – the worst off might simply desire less:

> A thoroughly deprived person, leading a very reduced life, might not appear too badly off in terms of the mental metric of desire and its fulfilment, if the hardship is accepted with non-grumbling resignation. In situations of longstanding deprivation, the victims do not go on grieving and lamenting all the time, and very often make great efforts to take pleasure in small mercies and to cut down personal desires to

> modest – 'realistic' – proportions . . . The extent of a person's deprivation, then, may not at all show up in the metric of desire-fulfilment, even though he or she may be quite unable to be adequately nourished, decently clothed, minimally educated and properly sheltered.
>
> (Sen, 1992a, p. 55)

The problem is at least as likely to apply over generations as it does within generations. A future generation living with fewer options is less likely to have preferences for other goods, as indeed did less well off generations in the past.

8 For an excellent discussion that argues for the view that 'the concept of the margin should play a marginal role in decisions about environmental goods', see Vatn (2000, p. 504).
9 I borrow the phrase from Steiner (1994, p. 169), which has an excellent discussion of the issues discussed here.
10 For a discussion see Cartwright (1999, pp. 1–19).
11 There are two senses in which one can talk of one good being a substitute for another. First, there is the technical sense: one good is a substitute for another if it serves the same purpose or end. Second, there is a broader economic sense of the term that is employed in welfare economics: one good can be a substitute for another not if it functions to achieve the same end, but if the end that it achieves is as good for a person's well-being as the end that would have been achieved by the other good. Two goods are substitute for another not in the sense that they do the same job, but rather in the sense that, as Steiner puts it, 'although they each do a different job, those two jobs are *just as good* as one another' (Steiner, 1994, p. 171, emphasis in the original). Substitution in the second economic sense allows for a much wider substitutability of goods than the first. It is this second sense that concerns us here. For a discussion see O'Neill et al. (forthcoming, ch. 11).
12 Compare Wolff (2003).
13 To say that different goods are not substitutable across different dimensions of welfare is not to say that there are no causal relations between goods in different dimensions: clearly there are relations, for example the quality of social relations and the capacity to control one's life have major effects on a person's physical health (Marmot, 2004). However, they do not substitute for say the nutritional conditions of health.
14 An earlier version of this chapter appeared as 'Sustainability: Ethics, Politics and the Environment' in J. O'Neill, I. Bateman and R. K. Turner (eds) *Environmental Ethics and Philosophy* (Aldershot: Edward Elgar, 2001). I would like to thank Edward Elgar for permission to reuse the material in this book. I would also like to acknowledge the support of a grant from the Arts and Humanities Research Board.

7 Wilderness, cultivation and appropriation

1 The full characterisation of wilderness in Section 2c of the United States Wilderness Act 1964 reads thus:

> A wilderness, in contrast with those areas where man and his own works dominate the landscape, is hereby recognized as an area where the earth and its community of life are untrammelled by man, where man himself is a visitor who does not remain. An area of wilderness is further defined to mean in this Act an area of undeveloped Federal land retaining its primeval character and influence, without permanent improvements or human habitation, which is protected and managed so as to preserve its natural conditions and which (1) generally appears to have been affected primarily by the forces of nature, with the imprint of man's work substantially unnoticeable; (2) has outstanding opportunities for solitude or a primitive and unconfined type of recreation; (3) has at least five thousand acres of land or is of sufficient size as to make practicable its preservation and use in an unimpaired condition; and (4) may

also contain ecological, geological, or other features of scientific, educational, scenic, or historical value.

2 For a response from a defender of wilderness see Rolston (1997).
3 For the debates about wilderness see Callicott and Nelson (1998); Cronon (1995). On exclusions see Guha (1997).
4 On the use-mention distinction see Hunter (1971, p. 10).
5 For a detailed examination of the relation between Locke's theory of property and justification of colonial expansion see Tully (1993a, 1993b ch. 5); Arneil (1996). The influence of the Lockean view is evident not only in appeal to it in subsequent legal claims, but also in economic theory. Consider for example the following from Smith:

> The whole of the savage nations which subsist by flocks have no notion of cultivating the ground. The only instance what has the appearance of an objection to this rule is the state of the North American Indians. They, tho they have no conception of flocks and herds, have nevertheless some notion of agriculture. Their women plant a few stalks of Indian corn at the back of their huts. But this can hardly be called agriculture. This corn does not make any considerable part of their food; it serves only as a seasoning or something to give a relish to their common food; the flesh of those animals they have caught in the chase . . . [I]n North America, again, where the age of hunters subsists, theft is not much regarded. As there is no property among them, the only injury that can be done is the depriving them of their game.
>
> (Smith, 1982, i.29, i.33, pp. 15–16)

6 The claims that cultivation gives *a* title to land is already in Aquinas who appeals in turn to Aristotle.

> If a particular piece of land be considered absolutely, it contains no reason why it should belong to one man more than to another, but if it be considered in respect of its adaptability to cultivation, and the unmolested use of the land, it has a certain commensuration to be the property of one and not of another man, as the Philosopher shows (Polit. ii, 2).
>
> (Aquinas, 1975, II.II, 57.3)

In Locke it becomes *the* title.

7 See Monbiot (1994, chs. 4–5). Consider, for example the Masai suffering from malnutrition and disease on scrubland bordering the Mkomazi Game Reserve from which they were forcibly evicted from land in 1988 (*Observer*, 6 April 1997, p. 12).
8 'Bushmen Fight to stay on in Last Botswana Haven', *The Times*, 5 April 1996, p. 11.
9 Richard Sylvan wrote his unpublished manuscript 'Dominant British Ideology' after a seminar on his work which was held near Colt Park in the Yorkshire Dales. My criticism of his comments here continues a discussion he provoked on that occasion. It was a discussion that was sadly cut short by his death.
10 Compare the remarks of farmers on conservation on Pevensey Levels in Burgess et al. (1995).
11 My thanks to Jacques Weber for putting this point to me.
12 The passages come in the context of Mill's rejection of certain landed rights:

> When the 'sacredness of property' is talked of, it should always be remembered, that any such sacredness does not belong in the same degree to landed property. No man made the land. It is the original inheritance of the whole species. Its appropriation is wholly a question of general expediency. When private property in land is not expedient, it is unjust. It is no hardship to any one, to be excluded from what others have produced: they were not bound to produce it for his use, and he loses nothing by not sharing in what otherwise would not have existed at all. But it is some hardship

> to be born into the world and to find all nature's gifts previously engrossed, and no place left for the new-comer. To reconcile people to this, after they have once admitted into their minds the idea that any moral rights belong to them as human beings, it will always be necessary to convince them that the exclusive appropriation is good for mankind on the whole, themselves included.
>
> (Mill, 1994, Book II, ch. 2, section 6)

13 It is echoed for example in the failed attempts to introduce rights of access in the late nineteenth and early twentieth centuries: 'no owner or occupier of uncultivated mountain or moor lands in Scotland shall be entitled to exclude any person from walking on such lands for the purposes of recreation or scientific or artistic study, or to molest him in so walking' (*Clause 2, Access to Mountains (Scotland) Bill 1884*).

14 Compare Bernard Williams' comments: 'a self-conscious concern for preserving nature is not itself a piece of nature: it is an expression of culture, indeed of a very local culture (though that of course does not mean it is not important)' (Williams, 1995a, p. 237).

15 An earlier version of this chapter appeared in *Philosophy and Geography* 5, 2002, pp. 35–50 (http://www.tandf.co.uk).

8 The good life below the snowline

1 For a good development of this point see Larmore (1987). For a response see O'Neill (1993a, ch. 7; 1995b; 1998a, chs. 2–3).

2 For a defence of cost-benefit analysis on these liberal grounds see Miller (1999). For discussions of the relationship between liberalism and environmentalism see Coglianese (1998); Wissenburg (1998); de-Shalit (2000).

3 This concern is also to be found in particular much mainstream environmental ethical reflection in philosophy which for similar reasons often eschews thick more specific ethical concepts in an attempt to create universal environmental ethic that is not relative to time and place. The tendency is reflected in particular in the use of Kantian and utilitarian approaches to both animal welfare and the environment. For those concerned with environmental and animal welfare issues, the need to launch criticism of practices from a standpoint that transcends the local is often taken to be of particular significance in that part of the enterprise is to show that there are non-human individuals and groups who have an ethical standing that is not recognised in most existing social practices (see, for example, Taylor, 1984, p. 159). The enterprise is a critical endeavour in defence of those who are rendered ethically invisible in the existing social worlds. The need for cosmopolitan language of universal thin concepts is also taken to be more acute in the environmental sphere in virtue of the fact environmental problems are global and hence require an ethical language that crosses cultures.

4 I think more adequate formulation of this distinction is between 'thin' and 'thick' uses of concepts: general thin concepts like 'good' have specific world-guided uses when employed with a substantive – for example where we talk of a 'good farmer'. (Indeed, if it is the case that 'good' is an attributive adjective, then thin uses of 'good' are elliptical – one can always ask – 'a good what?' (Austin, 1962, p. 69; cf. Geach, 1967). The thick–thin distinction in this sense is tied mainly to various debates in meta-ethics about the defensibility of centralism –the view that thin concepts, general normative concepts, are conceptually prior to and independent of thick concepts (Hurley, 1989, ch. 2) – and the defensibility of non-cognitivism – the view that the view that normative utterance express attitudes or preferences towards the world and are not assertions that are true or false of the world (for discussions of both sides of the debate see Blackburn, 1992, pp. 258–299; Dancy, 1995).

5 I touch here on only some parts of a defence. For fuller arguments see O'Neill (1993a, ch. 7; 1995b; 1998a, chs. 2–3).

6 I develop this distinction in more detail in O'Neill (1995b; 1998a, ch. 2).

7 For comments on notions of care in agriculture in the United States, see Nassauer (1988).
8 Typical comments were the following: 'Now they're trying to make them [farmers] park keepers.' 'There is a danger of people becoming non-farmers. Custodians' (O'Neill and Walsh, 2000).
9 A phrase used by a friend in a UK conservation agency to describe a district she was surveying for its plant varieties.
10 Rawls (1972, section 79): see the corresponding criticism of the ideal of private society.
11 This chapter draws on material that previously appeared in 'The Good Life Below the Snow-line: Pluralism, Community and Narrative', in F. Arler and I. Svennevig (eds) *Cross-Cultural Protection of Nature and the Environment* (Odense University Press, Odense, 1997). I would like to thank the publisher for permission to reuse that material here.

9 Deliberation, power and voice

1

> Civil society is composed of those more or less spontaneously emergent associations, organizations, and movements that, attuned to how societal problems resonate in private life spheres, distill and transmit such reactions in amplified form to the public sphere. The core of civil society compromises a network of associations that instutionalizes problem solving discourses on questions of general interest inside the framework of organized public spheres. These 'discursive designs' have an egalitarian, open form of communication around which they crystallize and to which they lend continuity and permanence.
>
> (Habermas, 1996, p. 367)

For discussions of this deliberative reading of civil society see Dryzek (2000, ch. 4) Young (2000, ch. 5).

2 Compare distinction between the 'defensive' and 'offensive' goals of new social movements in Cohen and Arato (1992, pp. 523–532)
3 Here for example is Tony Blair:

> The democratic impulse needs to be strengthened by finding new ways to enable citizens to share in decision making that affects them. For too long a false antithesis has been claimed between 'representative' and 'direct' democracy. The truth is that in a mature society representatives will make better decisions if they take full account of popular opinion and encourage public debate on the big decisions affecting people's lives. Across the West new democratic experiments are under way, from elected mayors to citizens' juries.
>
> (Blair, 1998, p. 15)

4 For example, Blair in his references to the anti-globalisation movement as an 'anarchist travelling circus' portrays protests as an attempt to disrupt democracy: 'such protests must not and will not disrupt the proper workings of democratic organisations'. 'Blair: Anarchists Will Not Stop Us', BBC online, 16 June 2001. The points are echoed by spokespersons for the World Bank. Hence the following responding to the cancellation of a meeting in Barcelona under pressure from the anti-globalisation movement:

> The intention of many of the groups who plan to converge on Barcelona is not to join the debate or to contribute constructively to the discussion, but to disrupt it . . . It is time to take a stand against this kind of threat to free discussion.
>
> (Caroline Anstey, quoted in 'Protest Fear Forces World Bank Switch', BBC online, 19 June 2001)

5 Compare for example the features that Cohen employs to characterise the ideal deliberative procedures (J. Cohen, 1989, pp. 21–23).

6 For example, it is possible for members of juries to change the set of options with which they are presented or to introduce topics not on the agenda for discussion. For example in the wet fens jury developed as part of the *VALSE* project, the jurors, quite insightfully, raised the topic of the role of supermarkets in determining the future of agriculture in the wet fens and suggested as new options responses to that role. For a discussion see Aldred and Jacobs (2000).

7 For a discussion of this point and an amplification of the other points about deliberation noted here see Fairclough (2000, ch. 5).

8 Thus, when a spokesperson for the World Bank states, 'our intention is to have dialogue, but it's impossible to do that with those who want to abolish you' (Caroline Anstey, quoted in 'New Look at Anti-Globalisation', Associated Press, 10 June 2001), the point also plays from the opposite side of the argument. One expresses one's attitude towards an institutional actor as a *persona non grata* through non-conversation.

9 The point comes in the context of a discussion of Elster's (1998c) account of deliberative democracy.

10 See for example Young's account of the role of rhetoric, storytelling and greeting in public discourse in Young (1996, 2000, ch. 2).

11 Habermas suggests that while 'it has the advantage of a medium of *unrestricted* communication', the informal public sphere is 'more vulnerable to the repressive and exclusionary effects of unequally distributed social power, structural violence, and systematically distorted communication than are the institutionalized public spheres of parliamentary bodies' (Habermas, 1996, pp. 307–308, emphasis in the original). My own view is that it this is less a matter of the degree of vulnerability and more a matter of the kinds of vulnerability to which each are open.

12 Moreover, this problem may be a necessary feature of deliberative institutions: the quality of deliberation requires smaller fora, the quality of representativeness requires larger fora and the two demands pull in the opposite direction.

13 For a classic statement of that account see Walzer (1984). Something of that view pervades Habermas' view that we need to defend spheres of communicative action from colonisation by instrumental action.

14 The environmental dimensions of the socialist calculation debates are often overlooked in the standard histories of the debates which tend to focus on the exchanges between Mises (1920, 1922) and Hayek (1935, 1949a, 1949b) on the one hand, and Lange (1936–1937) and Taylor (1929) on the other. (For accounts of these debates with different verdicts on who won see Buchanan, 1985 ch. 4; Lavoie, 1985; Shapiro, 1989; Blackburn, 1991; and Steele, 1992.) Environmental questions were not to the fore in those exchanges. However, arguments concerning environmental goods are to be found in both Mises' and Hayek's works on socialist calculation, in particular in their arguments with Neurath. For a discussion see O'Neill, 1999, 2002.

10 The rhetoric of deliberation

1 See the passage from Kant quoted in Chapter 1 footnote 9 on p. 198.

2 For an excellent discussion of Gorgias' position and Plato's response see Wardy (1996, chs. 2–3).

3 Young's defence of rhetoric, storytelling and greeting in public discourse is open to being read in this way (Young, 1996, 2000).

4 A classic statement of what I describe as the Kantian position is offered by Wolff, in his *In Defense of Anarchism*:

> The responsible man is not capricious or anarchic, for he does acknowledge himself bound by moral constraints. But he insists that he alone is the judge of those

> constraints. He may listen to the advice of others, but he makes it his own by determining for himself whether it is good advice. He may learn from others his moral obligations, but only in the sense that a mathematician learns from other mathematicians – namely by hearing from them arguments whose validity he recognizes even though he did not think of them himself. He does not learn in the sense that one learns from an explorer, by accepting as true his accounts of things one cannot see for oneself. Since the responsible man arrives at moral decisions which he expresses to himself in the form of imperatives, we may say that he gives laws to himself, or he is self-legislating. In short, he is *autonomous*.
>
> (Wolff, 1970, pp. 13–14, emphasis in the original)

I criticise this view of the autonomous thinker at length in O'Neill (1998a, ch. 7). How far Kant is himself a Kantian in this sense is I think a moot question. Unlike Wolff, Kant does recognise the fact that much of our belief is founded upon testimony and hence relies upon faith in others (Kant, 1987, pp. 468–469). However, his writings on enlightenment do tend to assume epistemic self-sufficiency as a condition of intellectual autonomy. It is also the case that Kant's positive remarks on testimony suggest that he takes individual observation to have epistemic priority over testimony – we are justified in trusting testimony since 'for one of those witnesses it was after all his own experience' (Kant, 1987, p. 469). In this sense Kant is an individualist. For critical discussion of this form of individualism see Coady (1992, chs. 1, 4 and passim).

5 Shapin's account is developed in detail (Shapin, 1994).

6 I have little quarrel with some of the more carefully stated claims of Barnes and Bloor (1982):

> Our equivalence postulate is that all beliefs are on par with one another with respect to the causes of their credibility. It is not that all beliefs are equally true or false, but that regardless of truth and falsity the fact of their credibility is to be seen as equally problematic.
>
> (Barnes and Bloor, 1982, p. 23)

I reject the claim that credibility is purely a matter of sociology. Likewise, much in Shapin's (1994) *A Social History of Truth* is quite consistent with the position developed here. My main disagreement lies in the book title and the justification offered for it in Chapter 1: the book is a social history of credibility. For an excellent discussion of the issues of realism and relativism in Shapin's book see Lipton (1998).

7 For a discussion of the issues involved here see the last chapter of Shapin (1994).

8 For a discussion see Sherman (1997, pp. 178–179 and passim).

9 For a discussion see Sherman (1997, pp. 141–158).

10 Although Seneca's view of the role of emotion in rhetoric is complex. He defends what might be called a two audience view – that there is the unadorned language of truth that the wise talk to each other, and the rhetorical language suitable for the masses:

> Language . . . which devotes its attention to truth ought to be plain and unadorned. [The] Popular style has nothing to do with truth. Its object is to sway a mass audience, to carry away unpracticed ears by the force of its onslaught.
>
> (Seneca, 1969, p. 83)

This two audience view is popular in the history of rhetoric. It is defended for example by Averroes for whom division of labour between logic and rhetoric is related to their respective audiences: – logic addresses those educated in the arts of demonstrative argument, rhetoric those who lack either the natural ability or the time to learn them (Averroes, 1973, ch. 1 and ch. 3, pp. 301 and 311) – and by Locke (1975, Book III, ch. X, para. 34, 508). Given this view, Seneca's concern with the role of emotions in rhetoric has more to do with its effect on the speaker than the audience. He allows the rhetorician to *feign* emotions such as anger in oratory in order to move others, but not to be himself moved by them (Seneca, 1995, II.17).

11 An earlier version of this chapter was published as 'The Rhetoric of Deliberation: Some Problems in Kantian theories of Deliberative Democracy', *Res Publica*, 8, 2002, pp. 249–268. The material is reused with the kind permission of Springer Science and Business Media. I would also like to acknowledge the support of a grant from the Leverhulme Trust.

11 Representing people, representing nature, representing the world

1

> the craft of Cartography attained such Perfection that the Map of a Single province covered the space of an entire City, and the Map of the Empire itself an entire Province. In the course of Time, these Extensive maps were found somehow wanting, and so the College of Cartographers evolved a Map of the Empire that was of the same Scale as the Empire and that coincided with it point for point. Less attentive to the Study of Cartography, succeeding Generations came to judge a map of such Magnitude cumbersome, and, not without Irreverence, they abandoned it to the Rigours of the sun and Rain. In the western Deserts, tattered Fragments of the Map are still to be found, Sheltering an occasional Beast or beggar; in the whole Nation, no other relic is left to the Disciple of Geography.
>
> (Borges, 1981, p. 131)

The idea of such a perfect map is one that Lewis Carroll anticipates in *Sylvie and Bruno Concluded*.

2 The point has more general relevance to questions around idealisation in models and theories. As Joan Robinson notes, 'A model which took account of all the variegation of reality would be no more use than a map at the scale of one to one' (Robinson, 1962, p. 33).

3 This is not to suggest that maps offer an adequate model of representation in social science or politics, although the analogy with political representation has been drawn, most notably by Mirabeau in a speech to the Estates of Provence in 1789: 'A representative body is for the nation what a map drawn to scale is for the physical configuration of its land; in part or in whole the copy must always have the same proportion as the original' (cited in Pitkin, 1967, p. 62).

4 The point touches on some more general issues around the use of qualitative research employing particular cases. Some of the historical background to the criticism of qualitative research is to be found in Mitchell (1983). For discussions of the methodological issues around case studies see Eckstein (1975); Platt (1992); Ragin (1992); Flyvbjerg (2001).

5 Fay (1975) remains a clear and concise statement of the problems.

6 For a classic clear statement of this point see Taylor (1971).

7 For typical uses of the phrase 'measuring rod' see Pigou (1952, p. 11) and Pearce et al. (1989, ch. 3).

8 For classic statements of the point see Harré (1970).

9 See Habermas (1984, 1996). The main point is stated pithily by Dryzek (1990):

> Communicative rationality clearly obtains to the degree social action is free from domination (the exercise of power), strategizing by the actors involved, and (self-) deception. Further, all actors should be equally and fully capable of making and questioning arguments (communicatively competent). There should be no restrictions on the participation of these competent actors. Under such conditions, the only remaining authority is that of a good argument, which can be advanced on behalf of

> the veracity of empirical description, the understanding, and, equally important, the validity of normative judgements.
>
> (Dryzek, 1990, p. 15)

10 The choice between individuals qua individuals and individuals qua members of groups was at the centre of the associationalist criticisms of liberal representative democracy. Typical is G. D. H. Cole:

> In the majority of associations, the nature of the relation is clear enough. The elected person . . . makes no pretension of substituting his personality for those of his constituents, or representing them except in relation to a quite narrow and clearly defined purpose or group of purposes which the association exists to fulfil . . . True representation . . . is always specific and functional, and never general and inclusive. What is represented is never man, the individual, but always certain purposes common to groups of individuals.
>
> (Cole, 1920, pp. 105–106)

For a discussion see Hirst (1990). Cole's defence of associationalism has relevance for recent problems in the politics of presence. Cole's point is this: no one can represent me in my individuality. Someone may be able to represent me under a specific description for a specific purpose. My shop steward may not represent me in my gender, my ethnicity, my conceptions of the good life, and so on, but she may represent me as an employee in a particular firm. Under other descriptions I may need distinct forms of representation.

11 There is a tension between some forms of a politics of presence with a politics of solidarity. Andrew Collier puts the point forcibly:

> the idea that each oppressed group ought to be fighting for itself rather than for the emancipation of all the oppressed, seems to me to be regrettable . . . If we ever get the Left off the ground as a serious political force again, let it be academics who fight for the railways and railway workers who fight for universities, men who specialise in feminism, and white who make it their priority to fight racism.
>
> (Collier, 1999, p. 49)

The concern is I think legitimate. However, a distinction needs to drawn between 'acting in solidarity with' and 'acting as a representative of'. The former can be done only by those who do *not* belong to the same group, but its legitimacy requires accountability to self-organised groups with whom solidarity is expressed. A politics of presence becomes a problem when the two are confused, when 'acting in solidarity with' and in this sense 'on behalf of' a particular group is taken to involve acting 'as a representative of' that group.

12

> No government by a democracy . . . either in its political acts or in the opinions, qualities and tone of mind which it fosters, ever did or could rise above mediocrity, except in so far as sovereign Many . . . let themselves be guided . . . by the counsel and influence of a more gifted and instructed One or Few.
>
> (Mill, 1974, p. 82)

13

> The history of all countries shows that the working class, exclusively by its own effort, is able to develop only trade union consciousness, i.e., the conviction that it is necessary to combine in unions, fight the employers, and strive to compel the government to pass necessary labour legislation, etc. The theory of socialism,

> however, grew out of the philosophic, historical, and economic theories elaborated by educated representatives of the propertied classes, by intellectuals.
>
> (Lenin, 1963, ch. II)

14 Thus it rules out the kind of economic model of democratic representation offered by Downs according to which the function of representatives is to discover and transmit to their party information about the desires of their constituents (Downs, 1957, pp. 88ff.).

15 Compare Gould:

> [O]ne cannot argue that *any* member of a group can equally well represent *all* members of a group. It would be odd indeed to think that Clarence Thomas could represent all African-Americans or that Margaret Thatcher could represent all women.
>
> (Gould, 1996, p. 184, emphases in the original)

16 Typical is Neurath's comment:

> Democracy is . . . a continual struggle between the expert who knows everything and makes decisions, and the common man with just enough information to hold the power of the expert in check. Our life is connected more and more with experts, but on the other hand, we are less prepared to accept other people's judgement, when making decisions . . . What is called democracy implies the rejection of experts in making decision, therefore democracy in Athens was based on lot.
>
> (Neurath, 1996, p. 251)

17 Thus Hobbes, for whom authorisation is all of representation, denies that 'inanimate' or 'non-rational beings' can authorise others to act for them (Hobbes 1968, ch. 16).

18

> There is wide agreement that the state should protect the interests of the future *in some degree* against the irrational discounting and of our preferences for ourselves over our descendants. The whole movement for conservation in the United States is based on this conviction. It is clear duty of Government, which is the trustee for unborn generations as well as for its present citizens, to watch over, and, if need be, by legislative enactment, to defend, the exhaustible natural resources of the country from rash and reckless spoliation.
>
> (Pigou, 1952, pp. 29–30, emphasis in the original)

19 The more general idea of representation through the internalisation of the interests and viewpoints of others is developed in Goodin (2000).

20 For Arendt's account of judgement and its development in her later writings, see the excellent interpretive essay by Beiner in Arendt (1982).

21 For discussions see Sagoff (1984); Callicott (1989, 1998); Rawles (1997); Jamieson (1998).

22 Interestingly, however, Habermas' own views on deliberation do not entail the possibility of convergence in environmental matters. Convergence through public deliberation is assumed only in the moral domain of 'the right', not the ethical/evaluative domain of 'the good' which may require rather self-interpretative discourses (Habermas, 1993, especially chs. 1, 5). There exists a divergence between the recent experiments in the practice of deliberative democracy in the environmental sphere and the Habermasian theory of deliberation. The experiments are precisely in a domain in which Habermas denies the possibility of convergence through public deliberation. For Habermas' own views on the limits of deliberation in the environmental sphere see for example Habermas (1993, ch. 2). My own view is that reasoned public deliberation is possible in the both

'ethical' and 'moral' domains as Habermas characterises them, but that given a plurality of goods, tragic conflicts can be unavoidable in both.

23 It is worth noting that the arguments just outlined are not incompatible with Habermas' own theoretical position as such. The last highlight the place of deliberative institutions in conditions of conflicts of power and interest in the actual world. The first raises conflicts in what Habermas would count as ethical and evaluative domains rather than the domain of morality.

24 An earlier version of this chapter appeared in *Environment and Planning C: Government and Policy*, 2001, 19, pp. 483–500. I would like to thank Pion Limited, London for permission to reuse the material. I would also like to acknowledge the support of the Arts and Humanities Research Board.

12 The political economy of deliberation

1 For a classic statement of that account see Walzer (1984).

2 Central to that coordination is the activity of the entrepreneur who is alert to new opportunities in the marketplace. Through the activities of the entrepreneur the market acts as a discovery procedure. The entrepreneur is faced with a second source of ignorance, a future that at the point of decision is unpredictable. Given the unpredictability of future knowledge and invention, future human wants are in principle unpredictable. Wants change with the invention and production of new objects for consumption. If one also accepts the claim often attributed to Popper that the progress of human knowledge is in principle unpredictable – if we could predict future knowledge, we would already have it (Popper 1944–1945) – then it follows that since human invention relies on the progress of knowledge, and wants are created by human invention, future human wants are also in principle unpredictable. Hence at any point in time, we are ignorant about the full range of future human wants. The market is presented as a discovery procedure in which different hypotheses about the future are embodied in entrepreneurial acts and tested in the marketplace (Hayek, 1978, pp. 179–190; cf. Kirzner, 1985). Through such acts market coordination is realised.

3 'The most persistent advocate of . . . *in natura* calculation is, significantly, Dr. Otto Neurath, the protagonist of modern "physicalism" and "objectivism"' (Hayek, 1942–1944, p. 170).

4 I discuss Neurath's debate with Hayek in more detail in O'Neill (2004). See also O'Neill (1999, 2003a, 2003b).

5 Wainwright (1994) examines the parallels between socialist traditions and Hayek.

6 Hayek puts the point thus: '[I]t may . . . prove to be far the most difficult and not the least important task for human reason rationally to comprehend its own limitations' (Hayek, 1942–1944, p. 162). The claim echoes one that Neurath earlier made and was at the centre of his own views of socialist planning: 'Rationalism sees its chief triumph in the clear recognition of the limits of actual insight' (Neurath, 1913, p. 8). The parallels between their views were noted by Neurath (1945) in his response to Hayek. For a discussion see O'Neill (2002, 2003a, 2004).

7 I would like to acknowledge the support of the Arts and Humanities Research Board and a Manchester University Hallsworth fellowship in writing this chapter, which draws on material that appeared in 'Neurath, Associationalism and Markets', *Economy and Society*, 32, 2003, pp. 184–206 (http://www.tandf.co.uk).

Bibliography

Aguilera-Klink, F. (1994) 'Some Notes on the Misuse of Classic Writings in Economics on the Subject of Common Property', *Ecological Economics*, 9, pp. 221–228.

Aldred, J. (1996) 'Existence Value, Moral Commitments and In-kind Valuation', in J. Foster (ed.) *Environmental Economics: A Critique of Orthodox Policy*. London: Routledge.

Aldred, J. and Jacobs, M. (1997) *Citizens and Wetlands: Report of the Ely Citizens' Jury*. Lancaster, UK: Centre for the Study of Environmental Change, Lancaster University.

Aldred, J. and Jacobs, M. (2000) 'Citizens and Wetlands: Evaluating the Ely Citizens' Jury', *Ecological Economics*, 34, pp. 217–232.

Alveres, C. and Billorey, R. (1988) *Damming the Narmada*. Penang: Third World Network.

Anderson, D. and Grove, R. (1987) 'The Scramble for Eden: Past, Present and Future in African Conservation', in D. Anderson and R. Grove (eds) *Conservation in Africa*. Cambridge: Cambridge University Press.

Anderson, E. (1990) 'The Ethical Limits of the Market', *Economics and Philosophy*, 6, pp. 179–206.

Anderson, E. (1993) *Value in Ethics and Economics*. Cambridge, MA: Harvard University Press.

Anderson, T. and Leal, D. (1991) *Free Market Environmentalism*. San Francisco, CA: Pacific Research Institute for Public Policy.

Aquinas, T. (1975) *Summa Theologica Vol. 37*. London: Eyre & Spottiswoode.

Arendt, H. (1968) 'Truth and Politics', in H. Arendt, *Between Past and Future*. Harmondsworth: Penguin.

Arendt, H. (1982) *Lectures on Kant's Political Philosophy*. Brighton: Harvester.

Aristotle (1908) *Metaphysics*. Oxford: Clarendon.

Aristotle (1928) *Topics*, trans. W. Pickard-Cambridge. London: Oxford University Press.

Aristotle (1946) *Rhetoric*, trans. W. Roberts. Oxford: Clarendon.

Aristotle (1948) *Politics*, trans. E. Barker. Oxford: Clarendon.

Aristotle (1985) *Nicomachean Ethics*, trans. T. Irwin. Indianapolis, IN: Hackett.

Armour, A. (1995) 'The Citizens' Jury as Model of Public Participation: A Critical Evaluation', in O. Renn, T. Webler and P. Wiedemann (eds) *Fairness and Competence in Citizen Participation*. Dordrecht: Kluwer.

Arneil, B. (1996) 'The Wild Indian's Venison: Locke's Theory of Property and English Colonialism in America', *Political Studies*, 44, pp. 60–74.

Arneson, R. (1989) 'Equality and Equality of Opportunity for Welfare', *Philosophical Studies*, 55, pp. 77–93.

Arnold, R. (1996) 'Overcoming Ideology', in P. Brick and R. Cawley (eds) *A Wolf in the Garden: The Land Rights Movement and the New Environmental Debate*. Lanham, MD: Rowman & Littlefield.

Arrow, K. (1984) 'Limited Knowledge and Economic Analysis', in K. Arrow, *The Economics of Information*. Cambridge, MA: Harvard University Press.

Arrow, K. (1997) 'Invaluable Goods', *Journal of Economic Literature*, 35, pp. 557–565.

Auden, W. H. (1968) *Collected Longer Poems*. London: Faber.

Augustine (1952) *On Christian Doctrine*. Chicago, IL: University of Chicago Press.

Austin, J. L. (1962) *Sense and Sensibilia*. Oxford: Clarendon.

Averroes (1973) 'The Decisive Treatise Determining the Nature of the Connection between Religion and Philosophy', in A. Hyman and J. Walsh, *Philosophy in the Middle Ages*. Indianapolis, IN: Hackett.

Baden, J. and Noonan, D. (eds) (1998) *Managing the Commons*, 2nd edn. Bloomington, IN: Indiana University Press.

Barnes, B. and Bloor, D. (1982) 'Relativism, Rationalism and the Sociology of Knowledge', in M. Hollis (ed.) *Rationality and Relativism*. Oxford: Blackwell.

Barry, B. (1990) *Political Argument*, 2nd edn. New York: Harvester Wheatsheaf.

Barry, B. (1999) 'Sustainability and Intergenerational Justice', in A. Dobson (ed.) *Fairness and Futurity*. Oxford: Oxford University Press.

Bateman, I. and Willis, K. (eds) (1999) *Valuing Environmental Preferences*. Oxford: Clarendon.

Bava Mahalia (1994) 'Letter from a Tribal Village', *Lokayan Bulletin*, 11(2–3), Sept.–Dec.

Becker, G. (1976) *The Economic Approach to Human Behaviour*. Chicago, IL: University of Chicago Press.

Beckerman, W. (1994) '"Sustainable Development": Is It a Useful Concept?', *Environmental Values*, 3, pp. 191–209.

Beckerman, W. (1999) 'Sustainable Development and our Obligations to Future Generations', in A. Dobson (ed.) *Fairness and Futurity*. Oxford: Oxford University Press.

Beckerman, W. (2000) 'Review of J. Foster (ed.) *Valuing Nature? Ethics, Economics and the Environment*', *Environmental Values*, 9, pp. 122–124.

Beckerman, W. and Pasek, J. (1997) 'Plural Values and Environmental Protection', *Environmental Values*, 6, pp. 65–86.

Beckerman, W. and Pasek, J. (2001) *Justice, Posterity and the Environment*. Oxford: Oxford University Press.

Benhabib, S. (ed.) (1996) *Democracy and Difference*. Princeton, NJ: Princeton University Press.

Birch, A. (1972) *Representation*. London: Macmillan.

Birks, P. and McLeod, G. (trans.) (1987) *Justinian's Institutes*. London: Duckworth.

Bishop, R., Champ, P. and Mullarkey, D. (1995) 'Contingent Valuation', in D. Bromley (ed.) *A Handbook of Environmental Economics*. Oxford: Blackwell.

Blackburn, R. (1991) 'Fin de Siècle: Socialism after the Crash', *New Left Review*, 185, pp. 5–67.

Blackburn, S. (1992) 'Through Thick and Thin', *Proceedings of the Aristotelian Society*, 66, pp. 258–299.

Blair, A. (1998) *The Third Way: New Politics for the New Century*. London: Fabian Society.

Bockstael, N. (1995) 'Travel Cost Models', in D. Bromley (ed.) *A Handbook of Environmental Economics*. Oxford: Blackwell.

Bohman, J. (1996) *Public Deliberation*. Cambridge, MA: MIT Press.

Bohman, J. and Rehg, W. (eds) (1997) *Deliberative Democracy: Essays on Reason and Politics*. Cambridge, MA: MIT Press.

Borges, J. L. (1967) 'The Aleph', in J. L. Borges, *A Personal Anthology*. London: Jonathan Cape.

Borges, J. L. (1979) 'The Congress', in J. L. Borges, *The Book of Sand*. Harmondsworth: Penguin.

Borges, J. L. (1981) 'Of the Exactitude of Science', in J. L. Borges, *A Universal History of Infamy*. Harmondsworth: Penguin.

Brennan, G. and Buchanan, J. (1980) *The Power to Tax*. Cambridge: Cambridge University Press.

Brick, P. and Cawley, R. (eds) (1996) *A Wolf in the Garden: The Land Rights Movement and the New Environmental Debate*. Lanham, MD: Rowman & Littlefield.

Bromley, D. (1991) *Environment and Economy: Property Rights and Public Policy*. Oxford: Blackwell.

Bromley, D. (ed.) (1995) *A Handbook of Environmental Economics*. Oxford: Blackwell.

Broome, J. (1991) *Weighing Goods*. Oxford: Blackwell.

Buchanan, A. (1985) *Ethics, Efficiency and the Market*. Oxford: Clarendon.

Buchanan, J. (1969) 'External Diseconomies, Corrective Taxes and Market Structure', *American Economic Review*, 59, pp. 174–177.

Buchanan, J. (1972) 'Towards Analysis of Closed Behavioural Systems', in J. Buchanan and R. Tollison (eds) *Theory of Public Choice*. Ann Arbor, MI: University of Michigan Press.

Buchanan, J. (1975) *The Limits of Liberty*. Chicago, IL: University of Chicago Press.

Buchanan, J. (1978) 'From Private Preferences to Public Philosophy: The Development of Public Choice', in J. Buchanan (ed.) *The Economics of Politics*. London: Institute of Economic Affairs.

Buchanan, J. (1986a) *Liberty, Market and State*. Brighton: Wheatsheaf.

Buchanan, J. (1986b) 'Rights, Efficiency and Exchange: The Irrelevance of Transaction Costs', in J. Buchanan, *Liberty, Market and State*. Brighton: Wheatsheaf.

Buchanan, J. and Tullock, G. (1962) *The Calculus of Consent*. Ann Arbor, MI: University of Michigan Press.

Buchanan, J. and Wagner, R. (1977) *Democracy in Deficit: The Political Legacy of Lord Keynes*. New York: Academic Press.

Burgess, J., Limb, M. and Harrison, C. M. (1988) 'Exploring Environmental Values through the Medium of Small Groups', Parts 1 and 2, *Environment and Planning A*, 20, pp. 309–326 and 457–476.

Burgess, J., Clark, J. and Harrison, C. M. (1995) *Valuing Nature: What Lies Behind Responses to Contingent Valuation Surveys?* London: UCL Press.

Burke, E. (1774) 'Speech to the Electors of Bristol at the Conclusion of the Poll', in E. Burke (1899) *The Works*, vol.3. London: Nimmo.

Burke, E. (1792) 'Letter to Sir Hercules Langrishe', in E. Burke (1899) *The Works*, vol.4. London: Nimmo.

Burke, E. (1826) *The Works*, vol.V. London: Rivington.

Burnheim, J. (1985) *Is Democracy Possible?* Cambridge: Polity.

Burnyeat, M. (1994) 'Enthymeme: Aristotle on the Logic of Persuasion', in D. Furley and A. Nehamas (eds) *Philosophy and Rhetoric: Essays on Aristotle's Rhetoric*. Princeton, NJ: Princeton University Press.

Callicott, J. B. (1989) 'Animal Liberation: A Triangular Affair', in J. B. Callicott, *In Defense of the Land Ethic*. Albany, NY: State University of New York Press.

Callicott, J. B. (1998) '"Back Together Again" Again', *Environmental Values*, 7, pp. 461–475.

Callicott, J. B. and Nelson, M. (eds) (1998) *The Great Wilderness Debate*. Athens, GA: University of Georgia Press.

Campbell, B. (2005) 'Nature's and its Discontents in Nepal', *Conservation and Society*, 3, pp. 323–353.

Carroll, L. (1962) *The Annotated Snark*. New York: Simon & Schuster.

Cartwright, N. (1999) *This Dappled World*. Cambridge: Cambridge University Press.

Chambers, S. (1997) *Reasonable Democracy: Jurgen Habermas and the Politics of Discourse*. Ithaca, NY: Cornell University Press.

Chang, R. (1997) *Incommensurability, Incomparability and Practical Reason*. Cambridge, MA: Harvard University Press.

Choudhary, K. (2000) 'Development Dilemma: The Resettlement of Gir Maldheris', *Economic and Political Weekly*, 35(30), pp. 22–28.

Claro, E. (2003) 'Value Pluralism, Norms of Exchange and the Environment: The Acceptability of Compensation in the Siting of Waste Disposal Facilities', PhD thesis, University of Cambridge.

Clifford, S. and King, A. (eds) (1993) *Local Distinctiveness: Place, Particularity and Identity*. London: Common Ground.

Coady, C. (1992) *Testimony*. Oxford: Clarendon.

Coase, R. (1960) 'The Problem of Social Cost', *Journal of Law and Economics*, 3, pp. 1–22.

Cochrane, G. (1998) 'Involving Stakeholders in Practice', in *Strengthening Decision-Making for Sustainable Development Workshop*. ESRC Global Environmental Change Programme. Eynsham Hall, Oxford, 15–16 June.

Coglianese, C. (1998) 'Implications of Liberal Neutrality for Environmental Policy', *Environmental Ethics*, 20, pp. 41–59.

Cohen, G. A. (1989) 'On the Currency of Egalitarian Justice', *Ethics*, 99, pp. 906–944.

Cohen, J. (1989) 'Deliberation and Democratic Legitimacy', in A. Hamlin and P. Pettit (eds) *The Good Polity*. Oxford: Blackwell.

Cohen, J. and Arato, A. (1992) *Civil Society and Political Theory*. Cambridge, MA: MIT Press.

Cole, G. D. H. (1920) *Social Theory*. London: Methuen.

Collier, A. (1990) *Socialist Reasoning*. London: Pluto.

Collier, A. (1999) *Being and Worth*. London: Routledge.

Commons, J. (1934) *Institutional Economics: Its Place in Political Economy*. New York: Macmillan.

Coote, A. and Leneghan, J. (1997) *Citizens' Juries: Theory into Practice*. London: Institute for Public Policy Research.

Craig, D. (1990) *On the Crofter's Trail: In Search of the Clearance Highlanders*. London: Jonathan Cape.

Cronon, W. (ed.) (1995) *Uncommon Ground*. New York: Norton.

Cupitt, D. (1993) 'Nature and Culture', in N. Spurway (ed.) *Humanity, Environment and God*. Oxford: Blackwell.

Dales, J. H. (1968) *Pollution, Property and Prices*. Toronto: University of Toronto Press.

Daly, H. E. (1995) 'On Wilfred Beckerman's Critique of Sustainable Development', *Environmental Values*, 4, pp. 49–55.

Dancy, J. (1995) 'In Defence of Thick Concepts', *Midwest Studies in Philosophy*, 20, pp. 263–279.

David, P. (1992) 'Intellectual Property Institutions and the Panda's Thumb', in *Centre for Economic Policy Research* no. 287. Stanford, CA: Stanford University.

Dearlove, J. (1989) 'Neoclassical Politics: Public Choice and Political Understanding', *Review of Political Economy*, 1, pp. 208–237.

Debord, G. (1977) *Society of the Spectacle*. Detroit, MI: Black and Red.

De Marchi, B., Funtowicz, S., Casio, S. and Munda, G. (2000) 'Combining Participative and Institutional Approaches with Multicriteria Evaluation', *Ecological Economics*, 34, pp. 267–282.

de-Shalit, A. (1995) *Why Posterity Matters: Environmental Policies and Future Generations*. London: Routledge.

de-Shalit, A. (2000) *The Environment between Theory and Practice*. Oxford: Oxford University Press.

Deutscher, I. (1996) *Stalin*. Harmondsworth: Penguin.

Dobson, A. (1996) 'Representative Democracy and the Environment', in W. Lafferty and J. Meadowcraft (eds) *Democracy and the Environment: Problems and Prospects*. Cheltenham, UK: Edward Elgar.

Dobson, A. (ed.) (1999) *Fairness and Futurity*. Oxford: Oxford University Press.

Downs, A. (1957) *An Economic Theory of Democracy*. New York: Harper Row.

Downs, A. (1967) *Inside Bureaucracy*. Boston, MA: Little Brown.

Dryzek, J. (1990) *Discursive Democracy: Politics, Policy and Political Science*. Cambridge: Cambridge University Press.

Dryzek, J. (1992) 'Ecology and Discursive Democracy: Beyond Liberal Capitalism and the Administrative State', *Capitalism, Nature and Socialism*, 3, pp. 18–42.

Dryzek, J. (1996) *Democracy in Capitalist Times*. Oxford: Oxford University Press.

Dryzek, J. (2000) *Deliberative Democracy and Beyond: Liberals, Critics, Contestations*. Oxford: Oxford University Press.

Duncan, C. (1992) 'Legal Protection for the Soil of England: The Spurious Context of Nineteenth Century "Progress"', *Agricultural History*, 46, pp. 75–94.

Dunleavy, P. (1991) *Democracy, Bureaucracy and Public Choice*. New York: Harvester Wheatsheaf.

Durkheim, E. (1964) *The Division of Labor in Society*, trans. G. Simpson. New York: The Free Press.

Dworkin, R. (1977) *Taking Rights Seriously*. London: Duckworth.

Dworkin, R. (1981) 'What is Equality? Part 1: Equality of Welfare; Part II: Equality of Resources', *Philosophy and Public Affairs*, 10, pp. 185–246 and 283–345.

Eckersley, R. (1999) 'The Discourse Ethic and the Problem of Representing Nature', *Environmental Politics*, 8, pp. 24–49.

Eckstein, H. (1975) 'Case Study and Theory in Political Science', in F. Greenstein and N. Polsby (eds) *Handbook of Political Science*, volume VII. Menlo Park, CA: Addison-Wesley.

Eliot, T. S. (1951) 'Tradition and Individual Talent', in T. S. Eliot, *Selected Essays*. London: Faber & Faber.

Elliot, R. (1982) 'Faking Nature', *Inquiry*, 25, pp. 81–93.

Elliot, R. (1997) *Faking Nature*. London: Routledge.

Elster, J. (1986) 'The Market and the Forum: Three Varieties of Political Theory', in J. Elster and A. Hylland (eds) *Foundations of Social Choice Theory*. Cambridge: Cambridge University Press.

Elster, J. (1998a) 'Deliberation of Constitution Making', in J. Elster (ed.) *Deliberative Democracy*. Cambridge: Cambridge University Press.

Elster, J. (1998b) Introduction, in J. Elster (ed.) *Deliberative Democracy*. Cambridge: Cambridge University Press.

Elster, J. (ed.) (1998c) *Deliberative Democracy*. Cambridge: Cambridge University Press.

Emerson, R. W. (1983) 'Nature: Addresses, and Lectures', in J. Porte (ed.) *Essays and Lectures*. New York: Viking.

English Nature (1993) *Position Statement on Sustainable Development*. Peterborough, UK: English Nature.

Environmental Resources Limited (1993) *The Valuation of Biodiversity in UK Forests*. Report for the Forestry Commission. London: Environmental Resources Limited.

Evernden, N. (1992) *The Social Creation of Nature*. Baltimore, MD: Johns Hopkins University Press.

Fairclough, N. (1999) 'Democracy and the Public Sphere in Critical Research on Discourse', in R. Wodak and C. Ludwig (eds) *Challenges in a Changing World: Issues in Critical Discourse Analysis*. Vienna: Passagen Verlag.

Fairclough, N. (2000) *New Labour, New Language?* London: Routledge.

Fay, B. (1975) *Social Theory and Political Practice*. London: Allen & Unwin.

Field, P. , Raiffa, H. and Susskind, L. (1996) 'Risk and Justice: Rethinking the Concept of Compensation', *Annals of the American Academy of Political and Social Science*, 545, pp. 156–164.

Fischoff, B. (1991) 'Value Elicitation: Is There Anything There?', *American Psychologist*, 46, pp. 835–847.

Fishkin, J. (1991) *Democracy and Participation: New Directions for Democratic Reform*. New Haven, CT: Yale University Press.

Flyvbjerg, B. (2001) *Making Social Science Matter: Why Social Enquiry Fails and How it Can Succeed Again*. Cambridge: Cambridge University Press.

Flyvbjerg, B., Holm, M. S. and Buhl, S. (2002) 'Underestimating Costs in Public Works Projects: Error or Lie?', *Journal of the American Planning Association*, 68, pp. 279–295.

Fortenbaugh, W. (1975) *Aristotle on Emotion*. London: Duckworth.

Freeman, A. M. (1995) 'Hedonic Pricing Methods', in D. Bromley (ed.) *A Handbook of Environmental Economics*. Oxford: Blackwell.

Frey, B., Oberholzer-Gee, F. and Eichenberger, R. (1996) 'The Old Lady Visits your Backyard: A Tale of Morals and Markets', *Journal of Political Economy*, 104(6), pp. 1297–1313.

Geach, P. (1967) 'Good and Evil', in P. Foot (ed.) *Theories of Ethics*. Oxford: Oxford University Press.

Geertz, C. (1973) *The Interpretation of Cultures*. New York: Basic Books.

Gellner, E. (1973) *Cause and Meaning in the Social Sciences*. London: Routledge & Kegan Paul.

Geras, N. (1986) 'The Controversy about Marx and Justice', in N. Geras, *Literature of Revolution*. London: Verso.

Gillespie, J. and Shepherd, P. (1995) *Establishing Criteria for Identifying Critical Natural Capital in the Terrestrial Environment*. Peterborough, UK: English Nature.

Golding, M. (1972) 'Obligations to Future Generations', *The Monist*, 56, pp. 85–99.

Goodin, R. (1992) *Green Political Theory*. Cambridge: Polity.

Goodin, R. (1996) 'Enfranchising the Earth, and its Alternatives', *Political Studies*, 44, pp. 835–849.

Goodin, R. (2000) 'Democratic Deliberation Within', *Philosophy and Public Affairs*, 29, pp. 81–109.

Gordon, J. (1994) *Canadian Round Tables and Other Mechanisms for Sustainable Development in Canada*. Luton, UK: Local Government Management Board.

Gould, C. (1996) 'Diversity and Democracy: Representing Difference', in S. Benhabib (ed.) *Democracy and Difference*. Princeton, NJ: Princeton University Press.

Griffin, J. (1986) *Well-Being*. Oxford: Clarendon.

Griffiths, T. and Colchester, M. (2000) *Indigenous Peoples, Forests and the World Bank: Policies and Practice*. Washington, DC: Forest Peoples Programme.

Grove, R. (1995) *Green Imperialism: Colonial Expansion, Tropical Island Edens and the Origins of Environmentalism 1600–1860*. Cambridge: Cambridge University Press.

Guha, R. (1997) 'The Authoritarian Biologist and the Arrogance of Anti-Humanism', *The Ecologist*, 27(1), pp. 14–20.

Guha, R. and Martinez-Alier, J. (1997) *Varieties of Environmentalism: Essays North and South*. London: Earthscan.

Guttman, A. and Thompson, D. (1996) *Democracy and Disagreement*. Cambridge, MA: Harvard University Press.

Habermas, J. (1975) *The Legitimation Crisis of Late Capitalism*, trans. T. McCarthy. Boston, MA: Beacon.

Habermas, J. (1984) *The Theory of Communicative Action; vol. I: Reason and the Rationalisation of Society*, trans. T. McCarthy. Boston, MA: Beacon.

Habermas, J. (1987) *The Philosophical Discourse of Modernity*, trans. F. Lawrence. Cambridge: Polity.

Habermas, J. (1993) *Justification and Application: Remarks on Discourse Ethics*, trans. C. Cronin. Cambridge, MA: MIT Press.

Habermas, J. (1996) *Between Facts and Norms*. Cambridge, MA: MIT Press.

Hahn, F. and Hollis, M. (1979) Introduction, in F. Hahn and M. Hollis (eds) *Philosophy and Economic Theory*. Oxford: Oxford University Press.

Hardin, G. (1998) 'The Tragedy of the Commons', in J. Baden and D. Noonan (eds) *Managing the Commons*, 2nd edn. Bloomington, IN: Indiana University Press.

Harré, R. (1970) *The Principles of Scientific Thinking*. London: Macmillan.

Havel, V. (1989) 'The Power of the Powerless', in V. Havel, *Living in Truth*. London: Faber & Faber.

Hayek, F. (1935) 'The Nature and History of the Problem', in F. Hayek ed., *Collectivist Economic Planning*, London: Routledge & Kegan Paul.

Hayek, F. (1941) 'The Counter-Revolution of Science' [*Economica*, 8, 1941, pp. 9–36, 119–150, 281–320], reprinted with revisions in F. Hayek (1979) *The Counter-Revolution of Science*. Indianapolis, IN: Liberty.

Hayek, F. (1942–1944) 'Scientism and the Study of Society' [*Economica*, 9, 1942, pp. 267–291; 10, 1943, pp. 34–63; 11, 1944, pp. 27–29], reprinted with revisions in F. Hayek (1979) *The Counter-Revolution of Science*. Indianapolis, IN: Liberty.

Hayek, F. (1949a) 'The Uses of Knowledge in Society', in F. Hayek, *Individualism and Economic Order*. London: Routledge & Kegan Paul.

Hayek, F. (1949b) 'Economics and Knowledge', in F. Hayek, *Individualism and Economic Order*. London: Routledge & Kegan Paul.

Hayek, F. (1978) 'Competition as a Discovery Procedure', in F. Hayek, *New Studies in Philosophy: Politics, Economics and the History of Ideas*. Chicago, IL: University of Chicago Press.

Hayek, F. (1979) *The Counter-Revolution of Science*. Indianapolis, IN: Liberty.

Hegel, G. (1952) *Philosophy of Right*. London: Oxford University Press.

Hirsch, F. (1977) *Social Limits to Growth*. London: Routledge & Kegan Paul.

Hirschman, A. (1970) *Exit, Voice and Loyalty*. Cambridge, MA: Harvard University Press.

Hirschman, A. (1977) *The Passions and the Interests*. Princeton, NJ: Princeton University Press.

Hirschman, A. (1985) *Shifting Involvements: Private Interests and Public Affairs*. Oxford: Blackwell.

Hirst, P. (ed.) (1989) *The Pluralist Theory of the State*. London: Routledge.

Hirst, P. (1990) *Representative Democracy and its Limits*. Oxford: Polity.

HM Government (1990) *This Common Inheritance*, Cm 1200. London: HMSO.

HM Government (1994) *UK Biodiversity Action Plan*, Cm 2428. London: HMSO.

Hobbes, T. (1968) *The Leviathan*. Harmondsworth: Penguin.

Hodgson, G. (1988) *Economics and Institutions*. Cambridge: Polity.

Hodgson, G. (1993) 'Institutional Economics: Surveying the "Old" and the "New"', *Metroeconomica*, 44, pp. 1–28.

Hodgson, G. (1999) *Economics and Utopia: Why the Learning Economy is Not the End of History*. London: Routledge.

Hohfeld, W. N. (1913) 'Some Fundamental Legal Conceptions as Applied in Judicial Reasoning', *Yale Law Journal*, 51, pp. 951–966.

Holland, A. (1995) 'The Assumptions of Cost-Benefit Analysis: A Philosopher's View', in K. Willis and J. Corkindale (eds) *Environmental Valuation: New Perspectives*. Wallingford, UK: CAB International.

Holland, A. (1997) 'Substitutability: Or, Why Strong Sustainability is Weak and Absurdly Strong Sustainability is Not Absurd', in J. Foster (ed.) *Valuing Nature?* London: Routledge.

Holland, A. and Rawles, K. (1994) *The Ethics of Conservation*. Report presented to the Countryside Council for Wales, Thingmount Series no.1. Lancaster, UK: Department of Philosophy, Lancaster University.

Hume, D. (1978) *A Treatise of Human Nature*. Oxford: Clarendon.

Hume, D. (1985) *Essays, Moral, Political and Literary*. Indianapolis, IN: Liberty.

Hunter, G. (1971) *Metalogic*. London: Macmillan.

Hurley, S. (1989) *Natural Reasons*. Oxford: Oxford University Press.

Huxley, A. (1929) 'Wordsworth in the Tropics', in A. Huxley, *Do What You Will*. London: Chatto & Windus.

Jacobs, M. (1991) *The Green Economy*. London: Pluto.

Jacobs, M. (1995) 'Sustainable Development, Capital Substitution and Economic Humility: A Response to Beckerman', *Environmental Values*, 4, pp. 57–68.

Jacobs, M. (1997) 'Environmental Valuations, Deliberative Democracy and Public Decision-Making Institutions', in J. Foster (ed.) *Valuing Nature?* London: Routledge.

Jamieson, D. (1998) 'Animal Liberation is an Environmental Ethic', *Environmental Values*, 7, pp. 41–58.

Jayal, N. G. (2001) 'Balancing Ecological and Political Values', in M. Humphrey (ed.) *Political Theory and Environment*. London: Frank Cass.

Joss, S. and Durant, J. (1995) *Public Participation in Science: The Role of Consensus Conferences in Europe*. London: Science Museum.

Kant, I. (1784a) 'Idea for a Universal History with a Cosmopolitan Purpose', in H. Reiss (ed.) (1991) *Political Writings*. Cambridge: Cambridge University Press.

Kant, I. (1784b) 'An Answer to the Question: "What is Enlightenment?"', in H. Reiss (ed.) (1991) *Political Writings*. Cambridge: Cambridge University Press.

Kant, I. (1793) 'On the Common Saying: "This may be true in theory, but it does not apply in practice"', in H. Reiss (ed.) (1991) *Political Writings*. Cambridge: Cambridge University Press.

Kant, I. (1795) 'Perpetual Peace: A Philosophical Sketch', in H. Reiss (ed.) (1991) *Political Writings*. Cambridge: Cambridge University Press.

Kant, I. (1933) *Critique of Pure Reason*, trans. N. Kemp Smith. London: Macmillan.
Kant, I. (1948) *Groundwork of the Metaphysics of Morals*, trans. H. Paton. London: Hutchinson.
Kant, I. (1964) *The Doctrine of Virtue: Part II of The Metaphysic of Morals*, trans. M. Gregor. New York: Harper & Row.
Kant, I. (1974) *Anthropology from a Pragmatic Point of View*, trans. M. Gregor. The Hague: Martinus Nijhoff.
Kant, I. (1987) *Critique of Judgement*, trans. W. Pluhar. Indianapolis, IN: Hackett.
Keat, R. (1981) 'Individualism and Community in Socialist Thought', in J. Mepham and D-H. Ruben (eds) *Issues in Marxist Philosophy 4*. Brighton: Harvester.
Keat, R. (1993) 'The Moral Boundaries of the Market', in C. Crouch and D. Marquand (eds) *Ethics and Markets*. Oxford: Blackwell.
Keat, R. (1997) 'Environmental Goods and Market Boundaries', in T. Haywood and J. O'Neill (eds) *Justice, Property and the Environment: Social and Legal Perspectives*. Aldershot, UK: Avebury.
Keat, R. (2000) *Cultural Goods and the Limits of the Market*. London: Macmillan.
Kenyon, W., Hanley, N. and Nevin, C. (2001) 'Citizens' Juries: An Aid to Environmental Valuation?', *Environmental Planning C*, 19, pp. 557–566.
Kirzner, I. (1985) *Discovery and the Capitalist Process*. Chicago, IL: University of Chicago Press.
Kunreuther, H. and Easterling, D. (1996) 'The Role of Compensation in Siting Hazardous Facilities', *Journal of Policy Analysis and Management*, 15(4), pp. 601–622.
Kymlicka, W. (1996) 'Three Forms of Group-Differentiated Citizenship in Canada', in S. Benhabib (ed.) *Democracy and Difference*. Princeton, NJ: Princeton University Press.
Lange, O. (1936–1937) 'On the Economic Theory of Socialism', in B. Lippincott (ed.) *On the Economic Theory of Socialism*. New York: McGraw-Hill.
Larmore, C. (1987) *Patterns of Moral Complexity*. Cambridge: Cambridge University Press.
Larrere, C. (1996) 'Ethics, Politics, Science, and the Environment: Concerning the Natural Contract', in J. B. Callicott and F. de Rocha (eds) *Earth Summit Ethics: Toward a Reconstructive Postmodern Philosophy of Environmental Education*. Albany, NY: State University of New York Press.
Lavoie, D. (1985) *Rivalry and Central Planning: The Socialist Calculation Debate Reconsidered*. Cambridge: Cambridge University Press.
Leff, E. (2000) 'Sustainable Development in Developing Countries: Cultural Diversity and Environmental Rationality', in K. Lee, A. Holland and D. McNeill (eds) *Global Sustainable Development in the 21st Century*. Edinburgh: Edinburgh University Press.
Lenin, V. I. (1963) *What is To Be Done?* Oxford: Oxford University Press.
Leopold, A. S., Cain, S. A., Cottam, C. M., Gabrielson, I. N. and Kimball, T. L. (1963) *Wildlife Management in the National Parks*. Advisory Board on Wildlife Management, Report to the Secretary, 4 March. Washington, DC: US Department of the Interior.
Light, A. (2000) 'Public Goods, Future Generations, and Environmental Quality', in A. Anton, M. Fisk and N. Holmstom (eds) *Public Goods: A New Direction in Political Morality*. San Francisco, CA: Westview.
Lippincott, B. (ed.) (1956) *On the Economic Theory of Socialism*. New York: McGraw-Hill.
Lipton, P. (1998) 'The Epistemology of Testimony', *Studies in the History and Philosophy of Science*, 29(1), pp. 1–31.
Locke, J. (1975) *An Essay Concerning Human Understanding*. Oxford: Oxford University Press.

Locke, J. (1988) *Two Treatises of Government*, ed. P. Laslett. Cambridge: Cambridge University Press.

Loomis, J. B., Lockwood, M. and DeLacy, T. (1993) 'Some Empirical Evidence on Embedding Effects in Contingent Valuation of Forest Protection', *Journal of Environmental Economics and Management*, 25, pp. 45–55.

Lukács, G. (1968) *History and Class Consciousness*. London: Merlin.

Lund, K. (2006) 'Finding a Place in Nature: Intellectual and Local Knowledge in a Spanish Natural Park', *Conservation and Society*, 3, pp. 371–387.

Lyotard, J. (1984) *The Postmodern Condition: A Report of Knowledge*. Manchester: Manchester University Press.

Lyotard, J. (1992) *The Postmodern Explained*. Minneapolis, MN: University of Minnesota Press.

McCarthy, T. (1995) 'Enlightenment and the Idea of Public Reason', *European Journal of Philosophy*, 3, pp. 242–256.

MacIntyre, A. (1985) *After Virtue*, 2nd edn. London: Duckworth.

MacKenzie, J. (1988) *The Empire of Nature: Hunting, Conservation and British Imperialism*. Manchester: Manchester University Press.

McKirahan, R. (1994) *Philosophy before Socrates*. Indianapolis, IN: Hackett.

Macnaghten, P. , Grove-White, R., Jacobs, M. and Wynne, B. (1995) *Public Perceptions and Sustainability in Lancashire*. Lancaster, UK: Centre for the Study of Environmental Change, Lancaster University and Lancashire County Council.

Macpherson, C. B. (1973) *Democratic Theory: Essays in Retrieval*. Oxford: Clarendon.

Manin, B. (1987) 'On Legitimacy and Political Deliberation', *Political Theory*, 15, pp. 338–368.

Mansfield, C., Van Houtven, G. and Huber, J. (1998) *Compensating for Public Harms: Why Public Goods are Preferred to Money*. Working Paper. Durham, NC: Duke University.

Marmot, M. (2004) *The Status Syndrome: How Social Standing Affects our Health and Longevity*. London: Bloomsbury.

Martell, L. (1992) 'New Ideas of Socialism', *Economy and Society*, 21, pp 152–173.

Martinez-Alier, J. (1987) *Ecological Economics*. Oxford: Blackwell.

Martinez-Alier, J. (1994) 'Distributional Conflicts and International Environmental Policy on Carbon Dioxide Emissions and Agricultural Biodiversity', in J. van den Bergh and J. van der Straaten (eds) *Toward Sustainable Development*. Washington, DC: Island Press / International Society for Ecological Economics.

Martinez-Alier, J. (1997) 'The Merchandising of Biodiversity', in T. Hayward and J. O'Neill (eds) *Justice, Property and the Environment: Social and Legal Perspectives*. Aldershot, UK: Avebury.

Martinez-Alier, J. (2002) *The Environmentalism of the Poor*. Cheltenham, UK: Edward Elgar.

Martinez-Alier, J. and O'Connor, M. (1996) 'Ecological and Economic Distribution Conflicts', in R. Costanza, J. Martinez-Alier and O. Segura (eds) *Getting down to Earth: Practical Applications of Ecological Economics*. Washington, DC: Island Press / International Society for Ecological Economics.

Martinez-Alier, J., Munda, G. and O'Neill, J. (1998) 'Weak Comparability of Values as a Foundation for Ecological Economics', *Ecological Economics*, 26, pp. 277–286.

Martinez-Alier, J., Munda, G. and O'Neill, J. (1999) 'Commensurability and Compensability in Ecological Economics', in C. Spash and M. O'Connor (eds) *Valuation and Environment: Principles and Practices*. Aldershot, UK: Edward Elgar.

Martinez-Alier, J., Munda, G. and O'Neill, J. (2001) 'Theories and Methods in Ecological Economics: A Tentative Classification', in J. Cleveland, D. Stern and R. Constanza (eds) *The Economics of Nature and the Nature of Economics*. Aldershot, UK: Edward Elgar.

Marx, K. (1843) 'Critique of Hegel's Doctrine of the State', in L. Colletti (ed.) (1974) *Early Writings*. Harmondsworth: Penguin.

Marx, K. (1972) *Capital, III*. London: Lawrence & Wishart.

Marx, K. (1973) *Grundrisse*. Harmonsdworth: Penguin.

Marx, K. (1974) 'On the Jewish Question', in L. Colletti (ed.) *Early Writing*. Harmondsworth: Penguin.

Meikle, S. (1979) 'Aristotle and the Political Economy of the Polis', *Journal of Hellenic Studies*, 79, pp. 57–73.

Meikle, S. (1985) *Essentialism in the Thought of Karl Marx*. London: Duckworth.

Merton, R. (1968) *Social Theory and Social Structure*. New York: The Free Press.

Midgley, M. (1981) 'Trying out One's New Sword', in M. Midgley, *Heart and Mind*. New York: St Martin's Press.

Mill, J. S. (1848) *Principles of Political Economy*. Oxford: Oxford University Press.

Mill, J. S. (1947) *A System of Logic*. London: Longmans.

Mill, J. S. (1974) *On Liberty*, edited J. Gray and G. Smith. London: Routledge.

Mill, J. S. (1975) *Considerations on Representative Government*. London: Oxford University Press.

Mill, J. S. (1994) *Principles of Political Economy*. Oxford: Oxford University Press.

Miller, D. (1992) 'Deliberative Democracy and Social Choice', *Political Studies*, 40, pp. 54–67.

Miller, D. (1997) 'Equality and Justice', *Ratio* (new series), 10(3), pp. 222–237.

Miller, D. (1999) 'Social Justice and Environmental Goods', in A. Dobson (ed.) *Fairness and Futurity*. Oxford: Oxford University Press.

Mises, L. von (1920) 'Die Wirtshaftrechung im Sozialistischen Gemeinwesen', *Archiv für Sozialwissenschaften*, 47, translated as Mises 1935.

Mises, L. von (1922) *Die Gemeinwirtschaft: Untersuchungen über den Sozialismus*. Jena, Germany: Gustav Fischer, second edition translated as Mises 1981.

Mises, L. von (1935) 'Economic Calculation in the Socialist Commonwealth', in F. Hayek (ed.) *Collectivist Economic Planning*. London: Routledge & Kegan Paul.

Mises, L. von (1981) *Socialism*. Indianapolis, IN: Liberty.

Mitchell, J. C. (1983) 'Case and Situation Analysis', *The Sociological Review*, 31, pp. 187–211.

Monbiot, G. (1994) *No Man's Land*. London: Macmillan.

Mueller, D. (2003) *Public Choice III*. Cambridge: Cambridge University Press.

Muir, J. (1912) 'The Yosemite', in J. Muir (1992) *The Eight Wilderness Discovery Books*. London: Diadem.

Nagel, T. (1970) *The Possibility of Altruism*. Oxford: Oxford University Press.

Nassauer, J. (1988) 'The Aesthetics of Horticulture: Neatness as a Form of Care', *HortScience*, 23, pp. 973–977.

Nassauer, J. (1989) 'Agricultural Policy and Aesthetic Objectives', *Journal of Soil and Water Conservation*, 44, pp. 384–387.

Neurath, O. (1913) 'The Lost Wanderers of Descartes and the Auxiliary Motive', in O. Neurath, *Philosophical Papers*. Dordrecht: Reidel.

Neurath, O. (1920a) 'Total Socialisation', in T. Uebel and R. Cohen (2004) *Otto Neurath: Economic Writings*. Dordrecht: Kluwer.

Neurath, O. (1920b) ‘A System of Socialisation’, in T. Uebel and R. Cohen (2004) *Otto Neurath: Economic Writings*. Dordrecht: Kluwer.

Neurath, O. (1928) ‘Personal Life and Class Struggle’, in O. Neurath, *Empiricism and Sociology*. Dordrecht: Reidel.

Neurath, O. (1942) ‘International Planning for Freedom’, in O. Neurath, *Empiricism and Sociology*. Dordrecht: Reidel.

Neurath, O. (1943) ‘Planning or Managerial Revolution (Review of J. Burnham *The Managerial Revolution*)’, *The New Commonwealth*, 8(4), pp. 148–154.

Neurath, O. (1944) *Foundations of the Social Sciences*. Chicago, IL: University of Chicago Press.

Neurath, O. (1945) ‘Physicalism, Planning and the Social Sciences’, ‘Bricks Prepared for a Discussion v. Hayek’ and ‘Correspondence with F. A. Hayek’ *Nachlass*, 243, Haarlem.

Neurath, O. (1946) ‘The Orchestration of the Sciences by the Encyclopedism of Logical Empiricism’, in O. Neurath, *Philosophical Papers*. Dordrecht: Reidel.

Neurath, O. (1973) *Empiricism and Sociology*. Dordrecht: Reidel.

Neurath, O. (1983) *Philosophical Papers*. Dordrecht: Reidel.

Neurath, O. (1996) ‘Visual Education’, in E. Nemeth and F. Stadler (eds) *Encyclopedia and Utopia: The Life and Work of Otto Neurath (1882–1945)*. Dordrecht: Kluwer.

Niskanen, W. (1971) *Bureaucracy and Representative Government*. Chicago, IL: Aldine-Atherton.

Niskanen, W. (1973) *Bureaucracy: Servant or Master*. London: Institute of Economic Affairs.

NOAA (National Oceanic and Atmospheric Administration) (1990) *Oil Pollution Act of 1990: Proposed Regulations for Natural Resources Damage Assessments*. Washington, DC: US Chamber of Commerce.

Norman, R. (1997) ‘The Social Basis of Equality’, *Ratio* (new series), 10(3), pp. 238–252.

Nussbaum, M. (1990) ‘Aristotelian Social Democracy’, in R. B. Douglass, G. Mara and H. Richardson (eds) *Liberalism and the Good*. London: Routledge.

Nussbaum, M. (1994) *The Therapy of Desire*. Princeton, NJ: Princeton University Press.

Nussbaum, M. (2000) *Women and Human Development: The Capabilities Approach*. Cambridge: Cambridge University Press.

O’Connor, M. and Muir, E. (1995) ‘Endowment Effects in Competitive General Equilibrium: A Primer for Paretian Policy Analysts’, *Journal of Income Distribution*, 5, pp. 145–175.

Oksanen, M. (1997) ‘The Lockean Provisos and the Privatisation of Nature’, in T. Hayward and J. O’Neill (eds) *Justice, Property and the Environment: Social and Legal Perspectives*. Aldershot, UK: Avebury.

Olson, M. (1965) *The Logic of Collective Action*. Cambridge, MA: Harvard University Press.

Olson, M. (1982) *The Rise and Decline of Nations*. New Haven, CT: Yale University Press.

Olwig, K. (1995) ‘Reinventing Common Nature: Yosemite and Mt. Rushmore – A Meandering Tale of a Double Nature’, in W. Cronon (ed.) *Uncommon Ground: Toward Reinventing Nature*. New York: Norton.

O’Neill, J. (1990) ‘Property in Science and the Market’, *The Monist*, 73, pp. 601–620.

O’Neill, J. (1992) ‘Altruism, Egoism, and the Market’, *Philosophical Forum*, 23, pp. 278–288.

O’Neill, J. (1993a) *Ecology, Policy and Politics: Human Well-Being and the Natural World*. London: Routledge.

O’Neill, J. (1993b) ‘Future Generations: Present Harms’, *Philosophy*, 68, pp. 35–51.

O’Neill, J. (1993c) ‘Science, Wonder and the Lust of Eyes’, *Journal of Applied Philosophy*, 10, pp. 139–146.

O'Neill, J. (1995a) 'Intrinsic Evil, Truth and Authority', *Religious Studies*, 31, pp. 209–219.
O'Neill, J. (1995b) 'Polity, Economy, Neutrality', *Political Studies*, 43, pp. 414–431.
O'Neill, J. (1995c) 'I gotta use words when I talk to you', *History of Human Sciences*, 8, pp. 99–106.
O'Neill, J. (1995d) 'Essences and Markets', *The Monist*, 78, pp. 258–275.
O'Neill, J. (1996) 'Who Won the Socialist Calculation Debate?', *History of Political Thought*, 27, pp. 431–442.
O'Neill, J. (1997a) 'Value Pluralism, Incommensurability and Institutions', in J. Foster (ed.) *Valuing Nature*. London: Routledge.
O'Neill, J. (1997b) 'Cantona and Aquinas on Good and Evil', *Journal of Applied Philosophy*, 14, pp. 97–106.
O'Neill, J. (1997c) 'Thinking Naturally', *Radical Philosophy*, 83, pp. 36–40.
O'Neill, J. (1998a) *The Market: Ethics, Knowledge, and Politics*. London: Routledge.
O'Neill, J. (1998b) 'Self-Love, Self-Interest and the Rational Economic Agent', *Analyse & Kritik*, 20, pp. 184–204.
O'Neill, J. (1998c) 'Against Reductionist Explanations of Human Behaviour: Rational Choice and the Unified Social Science', *Proceedings of the Aristotelian Society, Supplementary Volume*, 72, pp. 173–188.
O'Neill, J. (1998d) 'Rhetoric, Science and Philosophy', *Philosophy of the Social Sciences*, 28, pp. 205–225.
O'Neill, J. (1999) 'Ecology, Socialism and Austrian Economics', in E. Nemeth and R. Heinrich (eds) *Otto Neurath: Rationalität, Planung, Vielfalt*. Vienna: Weiner Reihe.
O'Neill, J. (2001a) 'Chekov and the Egalitarian', *Ratio* (new series), 14(2), pp. 165–170.
O'Neill, J. (2001b) 'Meta-ethics', in D. Jamieson (ed.) *Blackwell Companion to Environmental Philosophy*. Oxford: Blackwell.
O'Neill, J. (2002) 'Socialist Calculation and Environmental Valuation: Money, Markets and Ecology', *Science and Society*, 66, pp. 137–151.
O'Neill, J. (2003a) 'Neurath, Associationalism and Markets', *Economy and Society*, 32, pp. 184–206.
O'Neill, J. (2003b) 'Unified Science as Political Philosophy', *Studies in History and Philosophy of Science*, 34, pp. 575–596.
O'Neill, J. (2004) 'Ecological Economics and the Politics of Knowledge: The Debate between Hayek and Neurath', *Cambridge Journal of Economics*, 28, pp. 431–447.
O'Neill, J. and Walsh, M. (2000) 'Landscape Conflicts: Preferences, Identities and Rights', *Landscape Ecology*, 15, pp. 281–289.
O'Neill, J., Holland, A. and Light, A. (forthcoming) *Environmental Values*. London: Routledge.
Orwell, G. (1996) *Homage to Catalonia*. Harmondsworth: Penguin.
Ostrom, E. (1990) *Governing the Commons*. Cambridge: Cambridge University Press.
Palmquist, R. (1991) 'Hedonic Methods', in J. Braden and C. Kolstad (eds) *Measuring the Demand for Environmental Improvement*. Amsterdam: Elsevier.
Parfit, D. (1984) *Reasons and Persons*. Oxford: Oxford University Press.
Parfit, D. (1997) 'Equality and Priority', *Ratio* (new series), 10(3), pp. 202–221.
Pearce, D. (1993) *Economic Values and the Natural World*. London: Earthscan.
Pearce, D. and Moran, D. (1995) *The Economic Value of Biodiversity*. London: Earthscan.
Pearce, D. and Turner, K. (1990) *Economics of Natural Resources and the Environment*. New York: Harvester Wheatsheaf.
Pearce, D., Markandya, A. and Barbier, E. (1989) *Blueprint for a Green Economy*. London: Earthscan.

Pennington, M. (2001) 'Environmental Markets versus Environmental Deliberation: A Hayekian Critique of Green Political Economy', *New Political Economy*, 6(2), pp. 171–190.

Phillips, A. (1995) *The Politics of Presence*. Oxford: Clarendon.

Phillips, A. (1997) 'Dealing with Difference: A Politics of Ideas or a Politics of Presence', in R. Goodin and P. Pettit (eds) *Contemporary Political Philosophy*. Oxford: Blackwell.

Pigou, A. (1952) *The Economics of Welfare*, 4th edn. London: Macmillan.

Pitkin, H. (1967) *The Concept of Representation*. Berkeley, CA: University of California Press.

Plamenatz, J. (1938) *Consent, Freedom and Political Obligation*. London: Oxford University Press.

Plato (1987) *Gorgias*, trans. D. Zeyl. Indianapolis, IN: Hackett.

Platt, J. (1992) '"Case Study" in American Methodological Thought', *Current Sociology*, 40, pp. 17–48.

Pocock, J. G. A. (1975) *The Machiavellian Moment*. Princeton, NJ: Princeton University Press.

Pocock, J. G. A. (1985) 'The Mobility of Property and the Rise of Eighteenth-Century Sociology', in J. G. A. Pocock, *Virtue, Commerce, and History*. Cambridge: Cambridge University Press.

Polanyi, K. (1957a) *The Great Transformation*. Boston, MA: Beacon.

Polanyi, K. (1957b) *Primitive, Archaic and Modern Economies*. Boston, MA: Beacon.

Polanyi, K. (1957c) 'Aristotle Discovers the Economy', in K. Polanyi, *Primitive, Archaic and Modern Economies*. Boston, MA: Beacon.

Popper, K. (1944–1945) 'The Poverty of Historicism' [*Economica*, 11, 1944, pp. 86–103, 119–137; 12, 1945, pp. 69–89], reprinted with revisions in K. Popper (1986) *The Poverty of Historicism*. London: Routledge.

Porter, T. (1995) *Trust in Numbers: The Pursuit of Objectivity in Science and Public Life*. Princeton, NJ: Princeton University Press.

Radin, M. (1996) *Contested Commodities*. Cambridge, MA: Harvard University Press.

Ragin, C. C. (1992) '"Casing" and the Process of Social Enquiry', in C. Ragin and H. Becker, *What is a Case: Exploring the Foundations of Social Enquiry*. Cambridge, Cambridge University Press.

Ramsey, F. (1928) 'A Mathematical Theory of Saving', *The Economic Journal*, 38, pp. 543–559.

Ravetz, J. (1973) *Scientific Knowledge and its Social Problems*. Harmondsworth: Penguin.

Rawles, K. (1997) 'Conservation and Animal Welfare', in T. Chappell (ed.) *The Philosophy of the Environment*. Edinburgh: Edinburgh University Press.

Rawls, J. (1972) *A Theory of Justice*. London: Oxford University Press.

Rawls, J. (1996) *Political Liberalism*. New York: Columbia University Press.

Raz, J. (1986) *The Morality of Freedom*. Oxford: Clarendon.

Rippe, K. P. and Schaber, P. (1999) 'Democracy and Environmental Decision-Making', *Environmental Values*, 8, pp. 75–88.

Robinson, J. (1962) *Essays in the Theory of Economic Growth*. London: Macmillan.

Rolston III, H. (1997) 'Nature for Real: Is Nature a Social Construct?', in T. D. J. Chappell (ed.) *Respecting Nature: Environmental Thinking in the Light of Philosophical Theory*. Edinburgh: University of Edinburgh Press.

Runte, A. (1987) *National Parks: The American Experience*, 2nd edn. Lincoln, NE: University of Nebraska Press.

Ryle, G. (1971a) 'The Thinking of Thoughts: What is "Le Penseur" Doing?', in G. Ryle, *Collected Papers Volume II*. London: Hutchinson.

Ryle, G. (1971b) 'Thinking and Reflecting', in G. Ryle, *Collected Papers Volume II*. London: Hutchinson.

Sagoff, M. (1984) 'Animal Liberation and Environmental Ethics: Bad Marriage, Quick Divorce', *Osgoode Hall Law Journal*, 22, pp. 297–307.

Sagoff, M. (1988) *The Economy of the Earth*. Cambridge: Cambridge University Press.

Samuels, W. (1972) 'Welfare Economics, Power and Property', reprinted in W. Samuels and A. Schmid (eds) (1981) *Law and Economics: An International Perspective*. Boston, MA: Martin Nijhoff.

Schecter, D. (1994) *Radical Theories*. Manchester: Manchester University Press.

Schmid, A. (1978) *Property, Power and Public Choice*. New York: Praeger.

Schumpeter, J. (1987) *Capitalism, Socialism and Democracy*. London: Unwin.

Sen, A. (1980) 'Equality of What?', in S. McMurrin (ed.) *Tanner Lectures on Human Values*. Cambridge: Cambridge University Press.

Sen, A. (1987) *On Ethics and Economics*. Oxford: Blackwell.

Sen, A. (1992a) *Inequality Reexamined*. Oxford: Clarendon.

Sen, A. (1992b) 'Well-Being and Capability', in M. Nussbaum and A. Sen (eds) *The Quality of Life*. Oxford: Clarendon.

Sen, A. (1997) *On Economic Inequality*, 2nd edn. Oxford: Clarendon.

Sen, A. (1999) *Development as Freedom*. Oxford: Oxford University Press.

Sen, A., Muellbauer, J., Kanbur, R., Hart, K. and Williams, B. (1987) *The Standard of Living*. Cambridge: Cambridge University Press.

Seneca (1969) *Letters from a Stoic*. Harmondsworth: Penguin.

Seneca (1995) 'On Anger', in J. Cooper and J. Procopé, *Seneca: Moral and Political Essays*. Cambridge: Cambridge University Press.

Shapin, S. (1994) *A Social History of Truth*. Chicago, IL: University of Chicago Press.

Shapin, S. (1995) 'Cordelia's Love: Credibility and the Social Studies of Science', *Perspectives on Science*, 3, pp. 255–275.

Shapiro, D. (1989) 'Reviving the Socialist Calculation Debate: A Defense of Hayek against Lange', *Social Philosophy and Policy*, 6, pp. 139–159.

Sherman, N. (1997) *Making a Necessity of Virtue*. Cambridge: Cambridge University Press.

Shiva, V. (1992) 'Recovering the Real Meaning of Sustainability', in D. Cooper and J. Palmer, *The Environment in Question*. London: Routledge.

Sidgwick, H. (1907) *The Method of Ethics*. London: Macmillan.

Simon, H. (1972) *Theories of Bounded Rationality, Decision and Organisation*. Amsterdam: North Holland.

Simon, H. (1979) 'From Substantive to Procedural Rationality', in F. Hahn and M. Hollis (eds) *Philosophy and Economic Theory*. Oxford: Oxford University Press.

Smith, A. (1981) *An Inquiry into the Nature and Causes of the Wealth of Nations*. Indianapolis, IN: Liberty.

Smith, A. (1982) *Lectures on Jurisprudence*. Indianapolis, IN: Liberty.

Smith, G. (2003) *Deliberative Democracy and the Environment*. London: Routledge.

Solow, R. M. (1974) 'The Economics of Resources or the Resources of Economics', *American Economic Review*, 64, pp. 1–14.

Soper, K. (1996) *What is Nature?* Oxford: Blackwell.

Sprat, T. (1667) *The History of the Royal-Society of London*. London: by T. R. for J. Martyn and J. Allestry; republished 1959, London: Routledge.

Steele, D. (1992) *From Marx to Mises: Post-Capitalist Society and the Challenge of Economic Calculation*. La Salle, IL: Open Court.

Steiner, H. (1994) *An Essay on Rights*. Blackwell: Oxford.

Sunstein, C. (1997) 'Preferences and Politics', in R. Goodin and P. Pettit (eds) *Contemporary Political Philosophy*. Oxford: Blackwell.

Sylvan, R. (unpublished) 'Dominant British Ideology'.

Tawney, R. (1964) *Equality*. London: Unwin.

Taylor, C. (1971) 'Interpretation and the Science of Man', *Review of Metaphysics*, 25, pp. 1–45.

Taylor, C. (1992) *Multiculturalism and 'The Politics of Recognition'*. Princeton, NJ: Princeton University Press.

Taylor, F. (1929) 'The Guidance of Production in a Socialist State', in B. Lippincott (ed.) *On the Economic Theory of Socialism*. New York: McGraw-Hill.

Taylor, P. (1984) 'Are Humans Superior to Animals and Plants?' *Environmental Ethics*, 6, pp. 149–160.

Temkin, L. (1993) *Inequality*. Oxford: Oxford University Press.

Titmuss, R. (1970) *The Gift Relationship*. London: Allen & Unwin.

Tullock, G. (1970) *Private Wants, Public Means*. New York: Basic Books.

Tully, J. (1993a) 'Placing the "Two Treatises"', in N. Phillipson and Q. Skinner (eds) *Political Discourse in Early Modern Britain*. Cambridge: Cambridge University Press.

Tully, J. (1993b) *An Approach to Political Philosophy: Locke in Context*. Cambridge: Cambridge University Press.

Vatn, A. (2000) 'The Environment as Commodity', *Environmental Values*, 9, pp. 493–509.

Vatn, A. and Bromley, D. (1994) 'Choices without Prices without Apologies', *Journal of Environmental Economics and Management*, 26, pp. 129–148.

Veblen, T. (1919) *The Place of Science in Modern Civilisation and Other Essays*. New York: Huebsch.

Vogel, S. (1996) *Against Nature*. New York: State University of New York Press.

Wainwright, H. (1994) *Arguments for a New Left: Answering the Free Market Right*. Oxford: Blackwell.

Walsh, M. (1997) 'The View from the Farm: Farmers and Agri-environmental Schemes in the Yorkshire Dales', *The North West Geographer*, 1, pp. 17–28.

Walsh, M., Shackley, S. and Grove-White, R. (1996) *Fields Apart? What Farmers Think of Nature Conservation Schemes in the Yorkshire Dales. A Report for English Nature and the Yorkshire Dales National Park Authority*. Lancaster, UK: Centre for the Study of Environmental Change, Lancaster University.

Walzer, M. (1983) *Spheres of Justice*. London: Martin Robertson.

Walzer, M. (1984) 'Liberalism and the Art of Separation' *Political Theory*, 12, pp. 315–330.

Walzer, M. (1994) *Thick and Thin*. Notre Dame, IN: University of Notre Dame Press.

Wardy, R. (1996) *The Birth of Rhetoric: Gorgias, Plato and their Successors*. London: Routledge.

Weidner, H. (1993) 'Mediation as a Policy Instrument for Resolving Environmental Disputes – with Special Reference to Germany', *Veroffentlichung der Abteilung 'Normbildung und Umwelt' des Forschungsschwerpunkts Technik, Arbeit, Umwelt des Wissenschaftzentrums Berlin für Sozialforschung*. Berlin.

Weil, S. (1952) *The Need for Roots*. London: Routledge & Kegan Paul.

Wiggins, D. (1998) 'The Claims of Need', in D. Wiggins, *Needs, Values, Truth*, 3rd edn. Oxford: Clarendon.

Williams, B. (1985) *Ethics and the Limits of Philosophy*. London: Fontana.

Williams, B. (1995a) 'Must a Concern for the Environment Be Centred on Human Beings', in B. Williams, *Making Sense of Humanity*. Cambridge: Cambridge University Press.

Williams, B. (1995b) 'Reply to Nussbuam', in J. Altham and B. Harrison (eds) *World, Mind, and Ethics*. Cambridge: Cambridge University Press.

Willis, K., Garrod, G. and Benson, J. (1995) *Wildlife Enhancement Scheme: A Contingent Valuation Study of Pevensey Levels*. Peterborough, UK: English Nature.

Wissenburg, M. (1998) *Green Liberalism: The Free and Green Society*. London: UCL Press.

Wolf, C. (1987) 'Market and Non-market Failure: Comparison and Assessment', *Journal of Public Policy*, 7, pp. 43–70.

Wolff, J. (2003) *The Message of Redistribution: Disadvantage, Public Policy and the Human Good*. London: Catalyst.

Wolff, R. P. (1970) *In Defence of Anarchism*. New York: Harper & Row.

Wood, A. (1995) 'Exploitation', *Social Philosophy and Policy*, 12, pp. 136–158.

World Commission on Environment and Development (1987) *Our Common Future*. London: Oxford University Press.

Wynne, B. (1982) *Rationality and Ritual: The Windscale Inquiry and Nuclear Decisions in Britain*. Chalfont St Giles, UK: British Society for the History of Science.

Yearley, S. (1989) 'Bog Standards: Science and Conservation at a Public Inquiry', *Social Studies of Science*, 19, pp. 421–438.

Yeo, S. (1987) 'Three Socialisms: Statism, Collectivism, Associationalism', in W. Outhwaite and M. Mulkay (eds) *Social Theory and Social Criticism: Essays for Tom Bottomore*. Oxford: Blackwell.

Young, I. (1996) 'Communication and the Other: Beyond Deliberative Democracy', in S. Benhabib (ed.) *Democracy and Difference*. Princeton, NJ: Princeton University Press.

Young, I. (2000) *Inclusion and Democracy*. Oxford: Oxford University Press.

Name index

Subject index